I0024052

Charles Frederick Holder

Living Lights

A Popular Account of Phosphorescent Animals and Vegetables

Charles Frederick Holder

Living Lights
A Popular Account of Phosphorescent Animals and Vegetables

ISBN/EAN: 9783744674645

Printed in Europe, USA, Canada, Australia, Japan

Cover: Foto ©berggeist007 / pixelio.de

More available books at **www.hansebooks.com**

LIVING LIGHTS

PYROSOMA AND DIVER.

(*See page* 84.)

A POPULAR ACCOUNT OF

PHOSPHORESCENT ANIMALS AND VEGETABLES

BY

CHARLES FREDERICK HOLDER

FELLOW OF THE NEW-YORK ACADEMY OF SCIENCES, ETC.; AUTHOR OF "ELEMENTS
OF ZOÖLOGY," "MARVELS OF ANIMAL LIFE," "THE IVORY
KING," "WONDER WINGS," ETC.

LONDON

SAMPSON LOW, MARSTON, SEARLE, AND RIVINGTON,

St. Dunstan's House

Fetter Lane, Fleet Street, E.C.

1887

[All rights reserved]

TO MY FATHER, THIS VOLUME IS GRATEFULLY INSCRIBED, IN
REMEMBRANCE OF DAYS PASSED AMONG THE LIVING LIGHTS
OF THE OUTER REEF.

PREFACE.

THE object of the present work is to interest young people in natural history by the presentation of an attractive — indeed, marvellous — phase of nature, and to encourage healthful outdoor observation, as well as habits of investigation.

The subject chosen for this work — that embracing the phenomenon of luminosity in animals, plants, and inorganic matter, and especially those that seem intended as illuminators of the ocean — is one which has ever possessed a fascination for the author.

During many years spent on Southern shores, in constant association with the most attractive features of marine life, the remembrance of the splendors of the night festivals of these wondrous ocean forms is most enduring. No fairy tale of human invention can relate to us more fascinating scenes than are realized in Nature's carnivals of the sea. Not only is the surface of the ocean, when lashed into foam by the tempest, luminous, but the greater depths, where the water is cold, near the freezing-point, and subject to pressure so great that instruments of glass are shattered and reduced to powder, abound in living lights.

And this abyssal region, covered by miles in depth of water, and which was formerly considered to be the most desolate region upon the globe, is inhabited by light-givers of marvellous beauty and brilliancy.

The little *Malacosteus*, with its gleams of yellow and green; *Stomias*, with sparkling side-lights; the dazzling effulgence of *Pyrosoma;* the comet-like glare of *Medusæ*, with their tints of many colors, — present a series of wonders which must excite the admiration of the most indifferent observer.

In the United States, there are ten thousand enrolled young naturalists, comprising the Agassiz Association. As one of a committee solicited to answer questions propounded by the young people, members of this association and of the Chautauqua Circle. I have often been surprised at the nature of the queries, which shows that this army of young observers includes many who are not merely collectors of curiosities, but are naturalists in the best sense. They are systematic inquirers, and working in the right direction to become scientists, should they continue.

It is to these young scientists, their unscientific elders, and the boys and girls in general who have not yet had their interest aroused in Nature's works, that this volume is addressed ; and if some information is conveyed, while appearing merely to entertain, one object of the author will have been accomplished.

The subject of phosphorescence is one which affords the widest field for investigators ; as, while the most careful descriptions of the light-emitting organs have been made, the actual cause of animal phosphorescence is unknown. Material for study is ever at hand ; the fire-fly courts attention at every summer door-yard, and the pools of beach and cove are illumined by ocean forms. Even the simplest experiments are of the greatest interest. I have read by the light of a luminous beetle, and have determined the time of night while holding my watch in the glare of ocean animals. Von Bibra wrote his description of the *Pyrosoma* by its own light ; the shark of Bennett illuminated his cabin like a chandelier ; photographs have been taken by the light of luminous beetles and by phosphorescent plates ; and probably the day is not distant when more important uses will be found for this wonderful light, which, in default of a better name, we term phosphorescence. It is found in the animal, vegetable, and the mineral kingdoms ; in life and in death ; in growth and in decay. It illumines, but does not appear to consume, and without perceptible heat exists where ordinary combustion is impossible.

From the nature of the subject, it is evident that illustrations of the phosphorescence of marine animals must be more or less conjectural; and those given, representing over fifty luminous forms, show as nearly as possible the probable effect produced. As this work is scientific only so far as to secure accuracy, some technical details have been omitted. To compensate in a measure, I have appended a fairly complete bibliography of most important monographs and papers on the subject, which may be of value to those who wish to pursue the subject in its technical relations.

To render the work as popular as possible, certain systematic portions necessary to the student are placed in an appendix, and referred to by number. The whole work is also thoroughly indexed.

While the chief feature of the volume embraces the phosphorescence of animals, it has been deemed advisable to include reference to luminous plants, minerals, and certain atmospheric phenomena, which, if not strictly comprehended under our title, will perhaps not be considered entirely foreign nor uninteresting in this connection.

It is my agreeable duty to acknowledge here the courtesy and kindly attentions received from M. Raphaël Dubois of the Zoölogical Society of France; Professor H. Filhol; Professor H. H. Giglioli, Director of the Zoölogical Institute of Florence, Italy; Professor Carlo Emery of the University of Bologna, Italy; and M. Zenger of Prague, Hungary, who generously forwarded for my use their most recent papers on the subject of phosphorescence.

I have also to name with thanks for similar favors Dr. Gunther, keeper of the British Museum, and acknowledge the value of contributions from the works of M. Quatrefages of the Institute of France.

C. F. H.

PASADENA, CAL., July, 1887.

CONTENTS.

xi

.

LIST OF ILLUSTRATIONS.

List of Illustrations.

List of Illustrations.

LIVING LIGHTS

SEA BOTTOM.
1,500 metres, or one quarter of mile in depth.

LIVING LIGHTS.

CHAPTER I

STARS OF THE SEA.

A MONG the many revelations of modern science, none
have a more absorbing interest than those relating to
the illumination of the deep sea. Until within a few years
the ocean has been a sealed book. The surface forms only
were known; and it was assumed that, owing to the enormous pressure, lack of sunlight, and consequent darkness,
Nature, at least in the abyssal depths, was at fault, and this
vast region was devoid of life and incapable of supporting it.

Recent investigations, however, have shown the reverse,
and that this great area, with its plateaux, its mountain
ranges, and its isolated, coral-capped peaks, whose valleys are
now known to lie miles in ocean depths, teems with living
forms, and, far from being the dismal realm we had supposed, is a region of surpassing wonder; which we may,
in fancy, term that lower firmament, where float sparkling,
gleaming constellations, meteor-like disks and globes with
trailing luminosity, single stars and *nebulæ* of living lights.

The phosphorescence of the sea is no new discovery, and
those who have visited the seashore at night must have

1

witnessed this phenomenon. The region of coves and beaches along the shores of Eastern Massachusetts, around Nahant particularly, is a favorable one for its full display. As the waves come rolling in upon the rocks, or upon the long, expansive shingle, in tidal measure, we see the foaming crest, seemingly igniting all along the line, more and more intense in brilliancy, when, with a roar, it breaks, masses of scintillating liquid upon the sands. We glide over the smooth portions of this sea, our boat leaving a golden train ; and every dip of oar, or the dash of some affrighted fish, creates an equally vivid display. Even when not disturbed, looking down into the calm, clear depths, the same phenomenon is witnessed. Pale, ghostly forms are seen here and there, moving slowly about, while the seeming silvery atoms suggest the *nebulæ* of this submarine sky. Deeper yet, the bottom shows weird splendors. The great kelps are bedecked with mystic lights, and gleam like diamond's flash from ledge and rock.

These wonderful exhibitions of submarine illumination are due to the presence of luminous creatures, or in some cases to large animals swimming through immense numbers of small phosphorescent bodies, so appearing as light-givers themselves.

In nearly every branch of the animal kingdom we shall find these *living lights;* some marvellously brilliant, others glowing with dim rays, and all contributing often to wondrous illumination, far-reaching or circumscribed.

If the ocean which contains these wondrous forms should suddenly become dry, we should find that its contour is very similar to that of the land. There would be hills, valleys, plains, mountains, and seeming river-beds where currents

LUMINOUS PROTOZOANS.

Noctiluca
(in milk).

Noct luca miliaris
(magnified 100 diameters).

N. miliaris
(slightly magnified).

LUMINOUS PORTION OF NOCTILUCA
(highly magnified).

have flowed; and so sharply are these defined, where growing atolls and reefs abound, one may stand — as I have often done upon those of the Florida reef — and drop a leaded line almost directly to the bottom in the clear blue waters.

This submarine scenery would not show the rough and jagged outlines which are a characteristic of terrestrial mountain ranges. Nearly all prominences in water at a considerable depth are well rounded off by a coating of fine ooze, formed of the minute and delicate shells of the *globerigina*, one of the lowest organized of animal life. These little creatures live upon the bottom, or in the watery space above, and the ooze which makes the sea-bottom, in great thickness, is almost entirely made up of the dead and cast-off shells of these microscopical creatures. The chalk cliff of Dover, England, — that white headland which has given the ancient name of Albion to the mother country,—is an upheaved mass of the same material, once found in the ocean bottom, now elevated by some geological change, and hardened into chalk, which it really is. What a surprising monument, erected by Nature's processes from the myriads of bodies of her most minute and most simply organized animals!

The familiar modern term "protoplasm" represents what is know to be the simplest form of life; scarcely more, seemingly, than a bit of jelly, without form, and we might say void of organization, for it is alive, and yet has no nerves, no organized vessels which we can perceive, but exists in our pools as the least organized animal known.

There is a species which belongs to one of the numerous kinds or groups of this the first and least perfect of the animal kingdom, which has also the great distinction of being the best known and most brilliant of marine light-

bearers. This is the *Noctiluca*, or, as its name implies, the
night-light. This little creature, but little more than visible
to the naked eye, is the largest of the so-called infusorians;
others of this group of animals requiring the aid of a
microscope to determine the form. It is but little more
in structure than the bit of protoplasm, or simplest organism or animal known. It looks when magnified — its
natural size being about that of a pin's head — much more
highly organized than the others, by being almost a complete
globe, and provided with a whip-like process or member.
It is also veined somewhat, and reminds one of a currant
or gooseberry. Now, it is often noticed that the smaller the
animal, the more numerous; indeed, also, the more numerous its progeny. We may well be prepared, then, to hear
that these minute creatures often swarm on the ocean
surface in myriad masses.

Fig. 1. of Plate I. represents the *Noctiluca* magnified one
hundred times. Fig. 2. of the same, represents the appearance of the creature when luminous, and only slightly
magnified. The long lash which extends from the side
is the locomotive organ. It is attached to the body near
what is supposed to be the mouth; though these creatures
are so simple that many kinds, just below in organization,
have no definite mouth nor stomach, but absorb food from
any surface of the body which comes in contact with it.

This infusorian and most potent of living lights, albeit
of extreme minuteness and simplicity as an organism, is
abundant in the ocean along the European shores, and is
often seen in our north-eastern waters, notably off Portland
harbor and along shore to Cape Ann. I have enjoyed the
privilege of witnessing the fullest glory of this little crea-

ture's effulgence. In our so-styled ocean firmament these living asteroids shine forth in those waters, and rival, if not excel, in light-giving any other known creature.

In the endeavor to study the mysterious lights, I spent considerable time on a rocky point which jutted out into the sea, at Ogunquit, Me., with my microscope at hand, as near as possible to the water; thus examining them while comparatively fresh from the sea. In taking up the little creatures, they assume a pear shape, from contraction, — the only evidence, seemingly, of life, but blazing with a flashing light over their entire surface. We had the advantage of having specimens fresh at hand, yet there are certain appliances indispensable for such work which we did not have, and, therefore, could not then perfect our dissections sufficiently to get satisfactory results. We must refer the reader, therefore, to the experiments detailed in the Appendix.[1]

In watching the light of the *Noctiluca*, we are reminded of the flash-light of a light-house, — the gleam appearing and disappearing with considerable regularity. It is difficult to trace the light to any particular portion of the body. In Plate I., Fig. III., is shown the supposed luminous organs, which would seem to show that there are luminous spots. Sometimes the light seems to pervade the entire body; again, to be in the outer skin or cuticle. When the light appears after an intermission, the spots referred to become luminous first, the light extending to the outer surface.

The conditions most favorable for respiration produce the greatest exhibitions of light; thus, if the water is constantly aërated, or disturbed so that the air has access, the gleam is intensified. If the animal is touched with the point of a needle, the light is quickly visible; and just before death it is continuously luminous, the phosphorescence disappearing just after dissolution. Experiments have shown that in a vacuum the light diminishes, — carbonic gas producing the same result. Humboldt refers to his luminous appearance after bathing in water abounding in *Noctilucæ;* and among the curious experiments might be mentioned one where print was read by a gobletful of these little creatures which rendered them living lamps, literally.

M. de Tessan, a French observer, has recorded a phenome-enon, which, I should judge, was due to *Noctilucæ,* with perhaps the additional light of other forms. The accompanying picture on plate II. was made from his description, showing the light, and people upon the shore endeavoring to read by it. He writes: "On the 10th of April, in the evening, the sea in the roadstead of Simonstown, Cape of Good Hope, presented an extraordinary phosphorescence of a most vivid character. At whatever points the phosphorescence was greatest, the water was colored on the surface as red as blood; and it contained such an immense quantity of little globules that it had the consistency of sirup. A bucket of water taken up at one of these points, and filtered through a piece of linen, left in the filter a mass of globules greater in volume than the water that had passed through; in other words, the globules constituted more than half of the whole quantity of sea water taken up in the bucket. Viewed under the magnifying-glass, these globules presented the appearance of little transparent and inflated bladders, having on their surface a black point surrounded with equally black radiating *striæ.* . . . The least agitation or slightest contact made them throw out a vivid greenish light."

PLATE II.

M. DE TESSAN READING BY LIGHT OF
PHOSPHORESCENT SEA.

As the waves washed in, M. de Tessan describes the light as appearing like the vivid flashes of lightning. "It lighted up the chamber that I and my companions occupied in the house of Mr. Ball, though it was situated more than fifty yards distant from the breakers. I even attempted to write by the light, but the flashes were of too short duration."

When a vessel is ploughing through masses of these animals, the effect is extremely brilliant. An American captain states that when his ship traversed a zone of these animals in the Indian Ocean, nearly thirty miles in extent, the light emitted by these myriads of fire-bodies, of which he estimated there were thirty thousand in a cubic foot of water, eclipsed the brightest stars; the milky way was but dimly seen; and as far as the eye could reach the water presented the appearance of a vast, gleaming sea of molten metal, of purest white. The sails, masts, and rigging cast weird shadows all about; flames sprang from the bow as the ship surged along, and great waves of living light spread out ahead, — a fascinating and appalling sight.

The enormous quantity of *Noctilucæ* in the water explains the intensity of the light. In experiments made at Bologne, one-seventh to one-half of a given amount of water taken up consisted of these minute light-givers, and Rymer Jones found thirty thousand in a cubic foot. According to Quatrefages, the light of *Noctilucæ* in full vigor is a clear blue; but, if the water is agitated, it becomes nearly, if not quite white, producing rich silvery gleams sprinkled with greenish and bluish spangles.

Regarding the intensity of the light, a tube fifteen millimetres in diameter, containing a bed of *Noctilucæ* at the surface twenty millimetres thick, emitted light sufficient to see the face of a watch and read the figures; and, if the little creatures were agitated, time could be ascertained at a distance of a foot. M. Quatrefages found that the most delicate thermometer was not affected by the light; and he assumes that it is not combustion from the fact that oxygen gas, when introduced, does not restore the light after it has disappeared at the death of the animal. His conclusion is, that the light is produced by the contracting of the interior mass of the body; and that the flashes, or scintillations, are due to the rupture and rapid contraction of the filaments of the interior. The fixed light he explains as resulting from the permanent contraction of the contractile tissues adhering to the inner surfaces of the general envelope. Giglioli is especially enthusiastic over the light of the *Noctilucæ* and other forms; and to show its general distribution he says that in fifty-five thousand marine miles traversed by the "Magenta," the Italian exploring-ship, in four hundred and thirty-nine days, phosphorescence was observed more than half of the time. He met *Noctilucæ* in the Bay of Naples, at Rio, in the Straits of Banca, while in the east coast of Asia; and at Port Jackson "the same milky uniform light was seen, without any green or bluish tint," and again at Valparaiso. He observed, including *Noctiluca miliaris*, three luminous forms, all differing in the color of their light. The one observed on the Asiatic coast emitted a green light, and is called by M. Giglioli, *N. homogenea*. The Pacific form, *N. pacifica*, has a whitish luminosity, and differs from the others materially in form and structure.

In many of the ports of tropical and semi-tropical America, it is the custom to bathe in the ocean at night, the warmth of the water rendering such recreation enjoyable. A gentleman newly arrived at one of the places on the Pacific coast proceeded at night to take a bath, and, upon rising from the water, was astonished and amazed to find that his entire body was luminous, seeming covered with a coating of light, which he found originated from innumerable minute phosphorescent animals, which clung to his garments, and changed the water all about to a golden hue.

A distinguished professor at Keil was, perhaps, the first to discover luminous microscopic animals.[2]

The largest of these minute creatures is about one-eighth of a line, the smallest from a forty-eighth to a ninety-sixth of a line in size.

Giglioli has made some interesting observations regarding the phosphorescence of the lowest class of animal life, the protozoans, and with his colleague, Professor de Fillipi, intends publishing the results of their observations.[3]

CHAPTER II.

THE METEORS OF THE SEA.

A S the rushing comets dim the brightest luminaries with their radiance, so the ocean meteors, the moving *medusæ*, seem to excel in the glory of their light. The sea-jellies are among the commonest forms of the sea-shore. In the summer months the silvery sands are strewn with their glassy disks; unattractive then, but, once launched and imbued with life, possessed of many beauties of form and color. They range in size from those almost invisible to the naked eye, to giants weighing, it is estimated, over a ton. Many have a complicated structure; yet, in nearly all, the solid parts of the animal rarely represents over five per cent of the whole; and in specimens of a familiar northern kind, *Aurelia*, 95.84 is water. Little opportunity for light in such a creature, one would say; yet the simple jellies are numbered among the chief illuminators of the upper region of the ocean. I have observed them in the Atlantic, the Pacific, and in the Gulf of Mexico, in waters of various degrees of temperature; but, perhaps, the finest exhibition of their phosphorescence was seen off Boon Island, on the coast of Maine. The ocean surface seemed fairly bespangled with these living gems, which appeared surrounded by a halo of light. Each tentacle seemed to glow with an intense

PLATE III.

white heat; and, at a short distance, the streamers resembled
delicate lace, wrought in curious designs. Peering into the
depths, they appeared everywhere, moving in all directions,
surrounded by the mysterious light whose office it is difficult
to conjecture.

The vast numbers of *medusæ*, and their importance as
light-givers, may be realized from the remarks of Giglioli,
who states that their light was seen from the "Magenta"
over an area of forty-four degrees of latitude, and for nearly
thirty consecutive days. During the day they sank into the
greater depths, at night rising to the surface, and appearing
like moderator lamps. With their long groups of tentacles
trailing behind as they pulsate through the ocean waters,
they readily suggest the title, "Meteors of the Sea."

With few exceptions, the sea-jellies are light-givers. The
giant *Cyanea*, — one of which was measured by Mrs. Agassiz,
and found to be nearly six feet in diameter, and to have
tentacles over one hundred feet in length — emits a pale,
greenish light; and, if the entire mass is luminous, it must
present a wondrous appearance as it moves through the
water, like a gigantic meteor. As large as this giant is,
weighing many hundred pounds, it is produced from a delicate little creature which would hardly be noticed by the
casual observer.

One of the commonest forms along the New-England coast
is a diminutive jelly,[4] seemingly blown in glass by some
skilful worker. As it moves gracefully along, it emits
a light of a deep aurelian blue, vast numbers imparting a
metallic glitter to the water.

On some calm night, about a rocky point where the
current flows silently along, myriads of these wondrous

forms may be seen passing in review. Peering down into
the depths from our boat, we may see a pretty, shapely
jelly-fish, called *Zygodactyla*, a golden *ignis fatuus* of the ocean
waters; the *Melicertus*, another of the same family, surrounded by a golden radiance; and a stately *Rhizostoma*,
which Giglioli observed in fresh or brackish water in Batavia,
emitting a fixed, bluish light; while *Zina*, *Coryne*, *Eucope*
and *Clytia*, and a host of other exceedingly pretty sea-jellies,
add to the glories of the scene.

The delicate *Thaumantius* (Plate III., Fig. 3) and Oceanea
are resplendent light-givers. The latter, according to Ehrenberg, being "surrounded by a shining crown," while *Pelagia*
illumines the deep sea by its mystic rays.

Although we have established a rule to refer the most
of the technical names, with the more scientific matter, to
the Appendix notes, we are yet inclined to retain in the
text, occasionally, some names which are especially attractive. Thousands of marine animals have no other name
but the generic ones given them by discoverers; but in
many instances they are pretty, and there is no reason
why they should not be used, as they must become the
common name of the object, as well as its technical
one.

Other known light-givers are recorded in the Appendix,[5]
—all forms of the greatest delicacy and beauty.

Of a brilliantly phosphorescent form,[6] Professor Alexander
Agassiz says, "When passing through shoals of these
medusæ, ranging in size from a pin's head to several inches
in length, the whole water becomes so wonderfully luminous
that an oar dipped in the water up to the handle can be
seen plainly on dark nights by the light so produced. The

LUMINOUS SEA-JELLY AND MOLLUSK.

Beroe forskii.   *Cranchia scabra.*

seat of the phosphorescence is confined to the locomotive rows; and so exceedingly sensitive are they, that the slightest shock is sufficient to make them visible by the light emitted from the eight phosphorescent plates."

Professor Agassiz also states that the *Lucernaria*[7], a handsome green sea-jelly, emits a peculiar bluish light of an exceedingly pale steel color. While all these forms are beautiful individually, their combined forces˜produce an array of splendors hardly to be described. Such pyrotechnic displays of Nature are best observed during the autumn, when the jellies are wrecked and stranded; the waves hurling them in, and grinding them up upon the rocks, which appear bathed in warm, lambent lights.

At Spouting Horn, on the New-England coast, this luminous water is forced through a small chimney or crevice in the rocks, with a reverberating roar; sending skyward a column of gleaming water, that breaks in mid-air and falls in golden spray. In drifting along in a boat at this time, every movement of the oar produces the most astonishing results. A slight splash is followed by a blaze of light. By having a companion keep up a continuous motion of the water, I have almost been able to read the print of a newspaper by the light of these disintegrated forms. One of the most striking displays of this phenomenon I have ever witnessed was at the little port of Ogunquit, Me.

Returning, one dark night, from an off-shore fishing excursion, I saw, as we approached the harbor, an irregular row of lights, apparently lanterns in the hands of friends. We hailed, and not until we were nearly in the surf were we undeceived. The rocks were lined with kelp; and, when the waves came in, the glowing, sparkling mass of *medusæ* caught

upon the weed, remaining, as the water left it suspended, a
blaze of light, until the next wave broke. My companion,
an old fisherman, had also been deceived by the lights; and
we drifted there for some time watching these strange
spectres appear and disappear.

The *medusæ* differ in their methods of illumination. The
Obelia, as a free-swimming disk, is non-luminous; but the
stem, or trophosome, out of which it is developed, has a fluc-
tuating light extending up and down its surface. In many
medusæ the light appears to be confined to the upper portion
of the umbrella, to the tentacles, and to the margin of the
disk; but if an oar is thrust through it, or a freshly stranded
jelly is torn and cut upon the sand, every portion seems to
become more or less luminous.[8]

The little jelly-like creatures called "comb-bearers," or
Ctenophores, are nearly all wonderfully phosphorescent. Instead of moving as do the ordinary jelly-fishes, they have
rows of comb-like paddles which move up and down in regular measure as they float along. In the daytime the little
fins gleam with gorgeous iridescent hues; while at night
they are brilliantly luminous, even the eggs and embryos of
some emitting light.

The *Beroë* (Plate IV., Fig. 1) is the most familiar, but the
Pleurobrachia is the most graceful. Drummond refers to
these forms in the following lines, —

"Shaped as bard's fancy shapes the small balloon,
   To bear some sylph or fay beyond the moon.
   From all her bands see lurid fringes play,
   That glance and sparkle in the solar ray
   With iridescent hues. Now round and round
   She whirls and twirls; now mounts, then sinks profound."

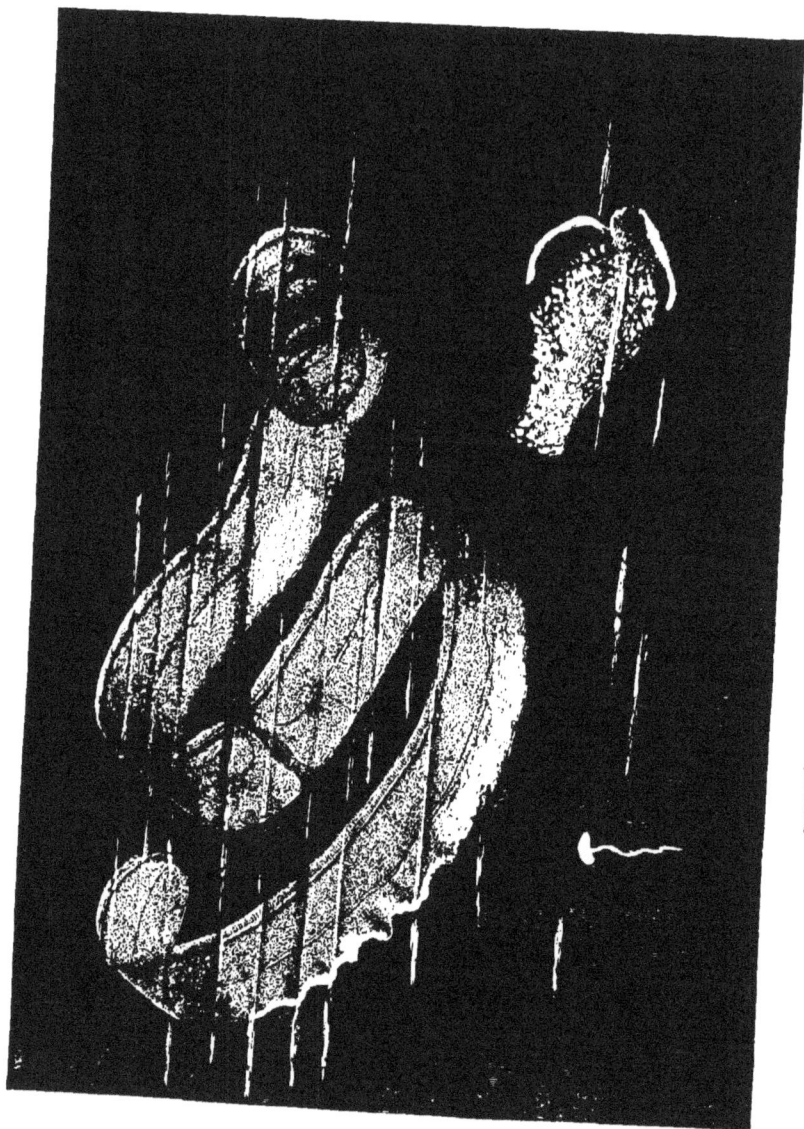

VENUS' GIRDLE.
(*Cestus veneris.*)

Phillirhoe.

So vast are the numbers of these and other light-givers in the northern seas, that the olive-green tints of the waters are due to them in the daytime. Mr. Scoresby, finding sixty-five of them in a cubic inch of water, summed up the interesting calculation, that, if eighty thousand persons had commenced at the beginning of the world (he refers to the popular, not geological, reckoning,) to count, they would barely at the present time have completed the enumeration of individuals of a single species found in a cubical mile.

One of the most remarkable of the Ctenophores is. the " Venus' girdle " (*Cestus veneris*), Plate V., Fig. 1. In shape it differs from all others of the class, as a comparison between it and the Beroë (Plate IV.), will show. It resembles in the daytime a silvery ribbon, or girdle, two or three feet in length, moving through the water by contractions of the body, rather than by the rows of combs that are found upon the edges. So delicate is this fragile creature, that it is almost impossible to remove it intact from the water. The mouth is in the centre, or equidistant between the ends; and on each side of it depends a short tentacle protruding from a sac. Opposite the mouth there is an otocyst, or sense-body. The combs, which are so conspicuous in other forms, are not so noticeable here, yet are well defined; and when moving along, and propelled by these gentle undulations, the *Cestus* is one of the most beautiful objects of the sea. At night this wonderful sea-ribbon develops a new charm, emitting, according to Giglioli, a reddish yellow light of singular brilliancy.

The Ctenophores, from their phosphorescence and great numbers, offer an interesting field for study. *Pleurobrachia* [9] may be found in myriads upon our eastern shores in the autumn. *Idya* [10] attracts immediate attention by its wondrous coloring, having a deep roseate hue. After death, its phosphorescence appears to be intensified, and much of the phosphorescent display is due to it. In nearly all the Cteno-phores the light is erratic, flash succeeding flash, and seeming, according to Giglioli, to reside along the zone covered by the vibrating *cilia*, or little paddles.

In the interesting group of animals known as *Physopho-ræ*,[11] or bubble-bearers, we find many light-givers of most remarkable form, in their structure reminding one of delicate objects in glass; and, according to Giglioli, all are more or less luminous. In the harbor of Gibraltar, he observed several beautiful forms, as *Abyla, Diphyes*, and *Eudoxia;* and in the Atlantic, in the latitude of Rio Janeiro, *Vogtia, Praia* (Plate VI., Fig. 2), *Abyla*, and *Eudoxia* were constantly encountered. These are all so fanciful in design, that they appear to be veritable fairy ships freighted with color-tints and gleams of light. Their luminosity is not scattered over the entire body as in many sea-jellies, but seems confined to fixed points, as in *Eucope*, a specimen of which, observed in the China Sea, seemed studded with brilliant emeralds, which appeared as marginal knobs at the base of the tentacles. In the Pacific, several species of *Diphyes* have been observed, their zooids [12] brilliantly phosphorescent; but the hydroids of this group, so far as known, are not luminous.

Many beautiful phosphorescent jellies can be observed, as we drift along, by using a small glass cylinder. With the finger pressed upon the top, lower the open end near the little creature, then remove the finger, when the jelly will be drawn into the improvised aquarium. If the night is dark, the play about its delicate form will be found a rare study.

Darwin refers to the beauties of the phosphorescent jellies

observed on one of his collecting-tours. He says, " While sailing a little south of the Plata on one very dark night, the sea presented a wonderful and most beautiful spectacle. There was a fresh breeze, and every part of the surface which during the day is seen as foam now glowed with a pale light. The vessel drove before her bows two billows of liquid phosphorus, and in her wake she was followed by a milky train. As far as the eye reached, the crest of every wave was bright; and the sky above the horizon, from the reflected glare of these livid flames, was not so utterly obscure as over the vault of the heavens. . . . Having used the net during one night, I allowed it to become partially dry; and having occasion, twelve hours afterward, to employ it again, I found the whole surface sparkling as brightly as when first taken out of the water. It does not appear probable, in this case, that the particles could have remained so long alive. On one occasion, having kept a sea-jelly of the genus *Diancea* till it was dead, the water in which it was placed became luminous. . . . Near Fernando Do Norhona, the sea gave out light in flashes. The appearance was very similar to that which might be expected from a large fish moving rapidly through a luminous fluid. To this cause the sailors attributed it; at the time, however, I entertained some doubts, on account of the frequency and rapidity of the flashes."

To Spallanzani is due the credit of first calling attention to the phosphorescence of the jelly-fishes or sea-jellies; he having observed it in the Mediterranean jelly, *Pelagia phosphorea*, which is luminous over its entire surface. He subsequently made some interesting experiments with *Aurelia phosphorea*, a jelly-fish similar to one on our coast, and came to the conclusion that the *light-emitting* organs lay in the arms, tentacles, and muscular zone of the body, and cavity of the stomach; the rest of the animal showing no luminosity. The light seemed to proceed from a viscous liquid, a secretion which oozed to the surface. One *Aurelia* that he squeezed in twenty-seven ounces of milk rendered the whole so luminous that a letter was read by the light, this being one of the first practical results of the investigation of marine phosphorescence. Humboldt experimented with *Aurelia aurita*, and, having placed it upon a tin plate, observed, that, whenever he struck it with another metal, the slightest vibration of the tin rendered the animal completely luminous. He also observed that it emitted a greater light when in a galvanic circuit.

CHAPTER III.

FIXED LUMINARIES OF THE SEA.

WE have examined and admired the movable and the moving luminaries of the ocean world, in the firmament of the deep, we may call it, — slowly moving stars of extreme minuteness, but great brilliancy, in one group, and the large orbs, more or less moving in erratic spheres, trailing in long lines of coruscating light, representing the lowest grand branches of the animal kingdom, the protozoans, and the vast colony of the sea-jellies, or *medusæ.*

We now come to the third chapter, which embraces those animals forming the grand branch of the animal kingdom which included in Cuvier's time the radiated animals.

All who have visited the New-England shores, or those beyond, farther north, or the warmer waters of our semi-tropical regions, have probably become acquainted with the soft and leathery forms, which, when seen undisturbed in the water, appear like flowers. For example, should we visit the delightful beaches and coves of Lynn, or Nahant, or Swampscott, the loved hunting-grounds of Agassiz and his disciples, we would see, on well-advised instruction from some one informed, what at first would suggest a moss pink in full bloom, nestled perhaps in groups, in crevice or open pool, among the crags or broken boulders. These are the com mon, and well nigh the only, representative of its family on our coast, within reach. Others there are, living in deeper water, within reach of a hand dredge, as work with such, in former years, well informed us. These are beautiful and very showy, like large asters and zinnias. But we dwell upon the in-shore one because it is always at hand and easily obtained, if you know where to look; and it well represents the characters of the group. Time was, when, forty years since, if some medical doctor of the town, or some of that ilk, did not have a sort of half knowledge of the creature, no one about did. But a few years before that, scientists in Europe were quarrelling over the question, Is it a vegetable, or an animal? Dr. Marsigli, a nobleman, asserted that such were vegetable, with further seeming good argument that the creatures looked like flowers and nothing else, therefore they must be flowers of the sea, notwithstanding that a poor, but educated Londoner, by the name of Ellis, demonstrated in good round science that they were animals. The striking case of mistaken identity, with the force of nobility, carried it. But Ellis lived to see his theory prevail.

Scarcely any in the whole range of Nature's objects are more surprising and more beautiful. The *Urticena nodosa* is a form found off our shores, which is luminous; the light being confined to its tentacles, and to the soft portion near the summit.

One of the most brilliant of this group of animals is the *Ilyanthus scoticus*, a kind usually found in ooze, the tentacles appearing at the surface, and gleaming brightly, like the rays of some fixed star. Even when brought up on the dredge, these animals emit a brilliant light.

Some of these sea-anemones are said to attach themselves

to the shells of hermit crabs; and, if luminous, we may imagine the spectacle of the gleaming, living light-house, moving about at the will of the little crustacean, possibly attracting prey to it instead of being the warning beacon that we might suppose. The anemones being, as a rule, fixed, one naturally likens this one to a light-ship which is drifting about away from its moorings.

The sea-anemones well repay examination and study, and thrive well in the aquarium, where their habits and development may be watched. As a rule, they are fastened to the rocks by a sucking disk. Some live in the mud; others float upon the surface, or are parasitic upon the great jelly-fishes. Some, as we have seen, ride about upon hermit crabs, or fasten themselves upon the claws of others; thus showing the greatest diversity in their life habits. The corals may be termed anemones which have the faculty of secreting or depositing lime, and among them are several which at times appear phosphorescent. The little cup-like *Caryophyllia* has been seen to emit a gleam of light, an idea of which is given in Plate III., Fig. 2.

The phosphorescence of reef-building corals has rarely been observed. Col. Nicolas Pike, our late consul to Mauritius, and an enthusiastic naturalist, informs me that he has witnessed the luminosity of their young. The account is so interesting an addition to the literature of the subject in general, that I give the colonel's letter entire : —

BROOKLYN, N.Y., December, 1886.

DEAR MR. HOLDER, — I remember on one occasion, when sailing on the Indian Ocean, the night was dark, but the crest of every wave glowed with light. As our vessel moved swiftly through the water, dashing the foamy waves on each side of her bows, she left bright streaks of light that reached far behind us. Every undulation of the water was lit up with scintillating points of light; and the ocean round us was so luminous, it would for splendor vie with the finest pyrotechnic display. So intense was the glow, the hull and sails of the vessel were illuminated by it; and, as I gazed at the glorious spectacle, I was filled with wonder and delight. The scene changed constantly, sometimes less brilliant than others, then again every rope in the ship was lit up; this, I presume, from the animals being more or less numerous. At the same time, darting in every direction, could be seen numerous fishes, making distinct streaks of light. Luminous spots from one to two inches in diameter were observed some distance under the water. These were *medusæ.* We captured many in our nets, and placed them in buckets of water on the deck, where they still continued to emit phosphorescent light. The grand scene lasted most of the night, and was faintly visible till dawn of day.

In the year 1868 or 1869, as I was dredging and collecting on the reefs near Port Louis, Mauritius, I met with one of the most singular sights it is possible to conceive. My Lascar crew gently moved my boat over the reefs, so that I could see any object in the water. The day was beautiful, not a cloud in the sky; but the bright sun shone down into the clear waters of the Indian Ocean, scarcely marked by a ripple on the surface. As the boat crossed over the shelving reef into deep water, what was my astonishment to find the depths alive with hundreds of millions of little creatures (which I supposed to be jelly-fish), actively moving about in the water, as far as the eye could penetrate. The little creatures, as they flitted about, emitted all the colors of the spectrum; the most brilliant diamond could not vie with the coruscations of light sent out by them. Such a scene must be witnessed to form any idea of its magnificence; the whole ocean was aglow with colored lights. I threw over my hand-net, and drew many thousands into the boat, which I placed in a large glass jar filled with sea-water, where I could examine them. They proved to be *young polyps of different species of the reef-building corals.* Those that I carried home were still luminous in the evening, and I thought I would pay a visit later, to see the effect at night; and it was truly great. The sea was one vast area of luminosity. The illumination extended for miles. Fishing-boats making for Port Louis harbor could be plainly seen a long way off by the phosphorescent light caused by the disturbance they made in passing through the immense shoal of coral polyps. The sight was curious and interesting; it had the appearance of an immense meteor coming directly down on our boat, as they were all heading for the entrance of the harbor where we lay. The bows of the fishing-boats made a great disturbance as they struck them, and the luminosity was most intense; but, as the waves were thrown off from the bows at a wide angle, the disturbance continued, and the colored light from the little creatures formed a long streak from behind the boat, representing the head of a comet with a long tail. Imagine twenty or thirty of these boats all heading in one direction, and you may form a faint idea of the scene. The polyps were not alone, but larger animals were darting and gyrating about, sending out vivid streaks of light.

The phosphorescent light of these polyps is probably the effect of a vital action ; it appears as a single spark, like that of various insects, and is repeated at short intervals.

In 1867 I passed through a belt of dark-colored water in a large stream. It had been observed from the masthead for sometime before we reached it : it proved to be a belt, of miles in extent, composed of animalcules. When taken up in a bucket, they gave out the strongest phosphorescent light I have ever witnessed. It required the highest power of my microscope to define them, and they were of many species new to me. Our steamer, a side-wheel vessel, made a great commotion as we passed for hours through this belt of living matter. These belts or patches, covering vast tracts of the ocean, are not uncommon. They are often seen in the Indian Ocean after severe storms and hurricanes. They vary in color. I have seen them of olive green, of a yellowish tint, and often a dark blue. Once, after a spent hurricane at Mauritius, I passed through a belt three miles wide, of a deep purple, so much so it could be seen a long way from shore.

Giglioli, the Italian naturalist, refers to the phosphorescence of madreporic polyps as being quite different from that of other forms. He observed on the coast of Sumatra and Batavia, that, when the bottom of his launch grounded

upon the polyps, a brilliant display of phosphorescence
followed.

Doubtless nearly all the Alcyonarian [13] corals are light-
givers, and of great importance in the illuminating economy
of the ocean.

The Alcyonarians include the sea-pens (*Pennatulidæ*),
and the sea-fans and the red coral of commerce (*Gorgonias*),
and may be briefly described as animals which, as a rule,
secrete a horny or calcareous stock, without the true divid-
ing septa that we see in the corals proper.

The Alcyonarians dredged by the "Challenger" were
almost invariably brilliantly luminous, making the dredge
appear as if red-hot coals were being taken up. The light
of the deep-sea forms was similar to that of those dredged on
shallow banks, where the phosphorescence is remarkably
brilliant; so that we can imagine the wondrous spectacle
presented in these little known regions. [14]

The sea-fans and plumes, known scientifically as Gorgonias, are extremely common upon the outer Florida reef, and
form the chief ornaments of these wonderful gardens of the
sea. We have drifted over them by day and night, peering
down into the depths, never wearying of the display. There
were two forms within diving distance on the reef, — one, a
rich yellow, reticulated fan; and the other, a vivid lilac. On
the yellow we often found a parasitic shell of almost the exact
hue of the Gorgonia; so like it, indeed, that it would not
have been noticed if the fan had not been closely examined,
— an interesting example of a protective resemblance.

At night these waters present a wondrous appearance,
gleams of light flashing from every direction. Even the
sand at the bottom seemed to give out fitful coruscations;

while pale, dim lights told of rare *medusæ*, — the phantoms of this world beneath the sea.

The gorgonias emit, as a rule, a light of a beautiful lilac hue; and in some localities the bottom of the ocean is covered with similar forms, all gleaming with this vivid phosphorescence. Imagine a cornfield covering hundreds of acres, the ripe ears emitting a fitful, vivid lilac light, through which dart various animals, — the birds of this submarine region, — their passage creating a blaze of another hue; and some idea can be formed of this scene that conjecture only can picture.

Sir Wyville Thompson states, that, when dredging in water nearly a mile deep off St. Vincent, they must have passed over an immense field of light-emitting gorgonias, as the trawls came up filled with a delicate form, "with a thin wire-like axis slightly twisted spirally, a small tuft of irregular rootlets at the base, and long exsert polyps. The stems, which were from eighteen inches to two feet in length, were coiled in great hanks round the trawl-beam, and entangled in masses in the net; and, as they showed a most vivid phosphorescence of a pale lilac color, their immense numbers suggested a wonderful state of things beneath."

Off our Eastern coast the little brush-like gorgonia, *Aca-nella*,[15] has been observed by Professor Verrill to emit a pale light when brought to the surface. The Gorgonias are all important light-givers. *Primnoa*,[16] a brush coral, and *Para-gorgia*[17] have become well known in late years by specimens brought up by the Gloucester fishermen on the Georges Banks. Even when dry and dead, they are extremely attractive; the *Primnoa* being richly tinted with pink, while the latter has a reddish hue.

If we could descend into these depths, we would find a veritable forest, with branches seeming on fire; many of the coral trees being from ten to fifteen feet in height, and equally as wide, forming lanes and open pathways through which the fishes pass, bathed in the wondrous light. That this is not imagination is shown by the branches brought up accidentally and by dredges; some limbs alone being four feet in length, and stout in proportion. Specimens of these forms can be seen in most of our museums of science; some sent by Dr. Holder from the Florida reef to the Museum of Natural History, in Central Park, are surprisingly beautiful.

More remarkable than the Gorgonias are the strange animals belonging to *Pennatulidæ*,[18] known popularly as the sea-pens (Plate VII., Fig. 2), from the resemblance of some to a quill pen, — an abnormally large one, it must be confessed. One of the most familiar forms is *Pennatula phosphorea.* When the animal is observed at night, and disturbed, it emits quite a brilliant light. In specimens observed at Oban by Professor Marchel, the more perfect females became vividly phosphorescent when the leaves were gently irritated. When the polyps were touched, they showed minute points of light, which appeared over the whole surface, in rapid, irregular coruscations.

If one of these living pens can produce so interesting a display, what must be the sight upon the bottom, where myriads of these curious forms abound, either fixed or moving!

It is not impossible that the light-emitting faculty of sea-pens is under control; at least, they have their periods of darkness and light. If a specimen which is not luminous is disturbed, as we have seen, it immediately becomes so. If the long axial stem is pinched, a seemingly protesting light appears on the lowest branchlets nearest the stem, quickly spreading, as if the polyps were igniting. When all those on a branch have become luminous, the light begins to appear on the next, and so on in succession until the whole glows brilliantly. Four-fifths of a second occur between the stimulation and the appearance of the light; so that in a sea-pen six and one-tenth inches in length, two seconds and a fifth were required for its complete illumination. By pinching the top or opposite end of the colony, the same phenomenon resulted, but reversed. If a polyp at the end of a branchlet was irritated, light immediately appeared, passed to its neighbor, and so on ; if a branch was touched at both ends, the light followed the act, and met in the centre.

These interesting experiments, which were made by Panceri, can be varied in many ways by those fortunate in securing a live sea-pen.[19]

The sea-pen *Pavonia*[20] is noted for its light-emitting properties ; and during the voyage of the English ship " Porcupine " the naturalists on board had many opportunities for observing its display. Sir Wyville Thompson, who was in charge, says, " Coming down the sound of Skye from Loch Torridon on our return, we dredged in about one hundred fathoms ; and the dredge came up tangled with the long pink stems of the singular sea-pen. Every one of these was embraced and strangled by the twining arms of an *Asteronyx*,[21] and the round soft bodies of the star-fishes hung from them like plump ripe fruit. The *Pavonariæ* were resplendent with a pale lilac phoshorescence, like the flame of Cyanogen gas ; not scintillating like the green light of some sea-stars,[22] but almost constant, sometimes flashing out at one point more

brightly, and then dying gradually into comparative dimness,
but always sufficiently bright to make every portion of a
stem caught in the tangles or sticking to the ropes distinctly
visible. From the number of specimens of sea-pens brought
up at one haul, we had evidently passed over a forest of
them. The stems were a meter long (over three feet)
fringed with hundreds of polyps."

When the ship " Venus " was lying off Simonstown, one
of their boats passed over a forest of sea-pens in shoal water,
which gave out a vivid light; while, where the ship lay at
anchor, other forms of phosphorescent animals illuminated
the ports so that the men lay in them and read by the
wondrous light on the darkest night.

The *Renilla*[23] is a rich purple species common on our
south-eastern borders.   Agassiz found it at Charleston, S.C.,
and says of its phosphorescence, that " it emitted a golden-
green light of wonderful softness."

Virgularia[24] is an attractive form ; and in certain portions
of the Patagonian coast they have been seen, when left by
the tide, emitting a light of great brilliancy.

Vertillum is an interesting genus, resembling a quill pen
in which the feathers have been curled or singed.   Its color
is a brilliant orange ; but in the darkness it develops a
phosphorescence of great beauty, and so penetrating that
a glass containing numbers of them has been used as a lamp
to read by, — an interesting example of one of the possible,
though not remarkably practical, uses of living lights to
mankind.

LUMINOUS STAR FISHES.
(*Brisinga elegans.*)
From 4,500 feet deep.

CHAPTER IV.

LUMINOUS ECHINODERMS.

IN the fourth grand branch of the animal kingdom, numerous creations are known which exhibit luminosity. The Echinoderms, as they are termed, are not well known to those who are not familiar with the seashore. To those who visit the marine beaches, one of the first objects that is met cast up by the tide, either fresh from its ocean bed among the rocks, or lying cast up high and dry amongst the vast masses of kelp, algæ, and other marine *débris*, is a sea-urchin, — so called for want of a better name, although the spines with which it is powerfully armed give good color to the nomenclature. The term Echinoderm is used to express all the kinds, as they have spines on the skin. As the arrangement of this division of Nature suggests, the creatures which are embraced here are next farther advanced in perfection of structure from the third, which includes the corals and sea-anemones. The animals are of most varied shape, exteriorly most unlike each other, yet internally possessing a structure each characteristic of the type. The sea-stars, forms quite as common as the sea-urchins which we first mentioned, are closely alike in structure, though so different in shape. Yet another form is seen in the celebrated trepang, which is dried, smoked, and sold to the Chinese for food, — a great luxury to them. Small species are found on our coast.

In some of these creatures the luminous property has been observed, — which usually surrounds the entire animal, — a pale light, rendering the object a beautiful one against the dark background of the ocean bottom. It is needless to say that the human eye has not penetrated these vast depths; but the ingenuity of the scientist has resulted in the invention of means by which the smallest as well as the largest of these strange creatures are dragged from their deep abode. Echinoderms are extremely numerous; on the Florida reefs we have often found it impossible to wade through considerable areas, where a kind of sea-urchin having long, slender black spines was so numerous as to pave the entire seabottom, and in certain localities in Long Island Sound we have seen the bottom fairly carpeted with star-fishes. It is not surprising, then, that the dredges of the "Challenger," "Porcupine," "Talisman," and other ships fitted out for scientific investigation, often came up loaded to overflowing with star-fishes, showing that the deep sea is equally populous with these living stars.

These deep-sea forms, especially of the *genera Asterias* and *Ophiura*,[25] are remarkable for their brilliancy, even when taken from their native element. When the bottom off the coast of Ireland was dredged by the "Challenger," an extraordinary number of luminous star-fishes were brought up from a depth of two-thirds of a mile. Several specimens are most noticeable for their brilliancy;[26] they appear as if burning internally with heat of great intensity. Even the mud about them was bespangled with luminous specks; and Sir Wyville Thompson says that in many instances every thing brought up in these waters was luminous. The light of one of the star-fishes was a brilliant green, and seemed to spring from the centre of the disk; flashing out now upon one arm, again upon another, or suddenly illumining the entire star in a brilliant aureola of phosphorescence.

This resplendent creature is especially common, according to Sir Wyville Thompson, off the coast of Stornaway and Shetland; and the nets, when hauled in, were often over-laden with masses of these gorgeous forms, which emitted a light of brilliant uranium green. Curiously enough, the young star-fishes exceeded the adults in the richness of their display. The gleams were not constant, but extremely erratic, appearing and re-appearing in a bewildering manner; and, according to the same naturalist, the most striking exhibitions were seen in very young ones.

The star-fishes known as Ophiuroids are among the most abundant of deap-sea forms. On the "Challenger," about several hundred species were brought up in the trawl from a depth of from half a mile to two and a half miles. In our own waters, two kinds [27] have been observed to emit a light of singular brilliancy.

Even more beautiful than these, as regards their luminosity, are the Brisingas,[28] one of which is shown with its light in Plate VIII., Fig. 1. This animal has nineteen long, snake-like arms, branching from a small central circular body. Its color in the daylight is a rich orange red; but at night, when taken from the dredge, it displays a vivid phosphorescence.

This attractive animal was first observed near Bergen, Norway, by Charles Abjördsen, who took a specimen in two hundred fathoms of water. Regarding it, he said, "it is a true *gloria maris*," and gave it the name of *Brisinga*, one of the jewels of the Goddess Freya.

The Brisingas have the faculty, common to many of their

allies, of casting their arms when touched; so that it is extremely difficult to take them intact. In lifting an *Astro-phyton*[29] from a branch of coral, we have had it drop into myriads of pieces ; so that there was a mimic rain of arms upon the bottom. This we found could be avoided by making the transfer under water, and, when the " basket-fish " was safely in the jar, killing it by the introduction of alcohol.

As to the cause of the light in the star-fishes, little is known. Quatrefages, after a careful examination of an Ophiuran, came to the conclusion that the light emitted was due to muscular contraction ; observing it arising between the plates of the arms and not on the disk, where, however, it has been seen since his observations were made. Professor P. Martin Duncan found upon examining a specimen, brought from the icy sea of North Smith's Sound, by Sir George Nares's expedition, that it had a delicate mucous envelope, which, he thought, in the young covered the plates and bases of the spines. In this filmy covering, he suggests, may be found the seat of the illuminating power.

CHAPTER V.

SUBTERRANEAN LIGHT-GIVERS.

IN wandering through the fields in early morning, we often see little heaps of newly disturbed earth, and occasionally catch glimpses of reddish or pink bodies quickly withdrawing into little tunnels in the sod. These are the earthworms, considered the humblest of all animals ; yet, as insignificant as they seem, they are among the most valuable aids to the agriculturist.

We may appreciate this by selecting a field at random in a good producing country, making a section down through the earth for several feet, when, if carefully done, we shall find innumerable tunnels formed by the worms, leading here, there, and everywhere. In fact, the upper crust of the earth is an endless maze of streets, lanes, and avenues. A naturalist has even attempted to calculate the number of these little workers, and has come to the conclusion that they average one hundred thousand to the acre; and in especially rich ground in New Zealand it was estimated that there were three hundred and forty-eight thousand, four hundred and eighty in a single acre. This vast body of worms is continually at work, boring this way and that, coming to the surface during the night, and retreating to greater depths during the day ; and it is at once evident that their tunnels constitute a system of irrigation and ventilation for the upper crust. In other words, rain, instead of running off, enters the holes, and so penetrates the earth, thus being held for a longer time. Air also finds its way below the surface; so that the homes of the little creatures constitute storehouses for moisture.

But this is a very small part of the work accomplished. The worms are in league with the farmer; are, in fact, his unappreciated assistants, upon whose endeavors depends much of the success of his crops. They are continually swallowing the earth, and depositing it at the surface, and working it over and over. If I should ask my young readers to estimate the quantity of earth brought to the surface in a single acre in a year, I fear they would not place the amount as high as Mr. Darwin, who states that the vegetable mould thus transported in some places amounts to ten tons an acre. Think of it! If your ten-acre farm is in one of these favored localities, these silent workers, say to the number of a million, have ploughed up about one hundred tons of earth for you, giving you a fine top dressing.

The worms not only carry all this material to the surface, but they drag vast quantities of leaves and other matter down, that serve to enrich the soil and render it capable of producing larger crops. They cover up seeds and other objects to a remarkable extent; and a flat rock set upon the ground will soon become buried, through their means. Some of the most interesting parts of Roman villas found in England have been, according to Darwin, preserved in this way; the worms undermining them, and gradually heaping soil over the walks and slabs, until finally, aided by other causes, they disappeared beneath the ground.

The earthworms of Australia attain a large size, — sometimes several feet in length, — and have been seen climbing trees. Some casts found in India are a foot in length. The worms evidently live in complete darkness; but it is known that at certain times, and under certain conditions, they are luminous: so that a state of things may exist under the ground of which we have no conception, and the tunnels of these little creatures may be brightly illumined. We have never been so fortunate as to observe their phosphorescence, but Dr. Phipson says, "I distinctly remember witnessing, when a child, the phosphorescence of the earthworm. The light appeared connected with the mucus that covered the animal's body." And other naturalists have observed the light under certain conditions.

If they possess this property to a greater extent than we are now aware of, it must be a fatal gift, as the sharp little eyes of the mole, though not remarkable for their powers of observation, would probably catch the faintest gleam. These animals are continually upon the forage; and their appetites can be imagined from an actual experiment, which showed that two moles devoured in nine days 341 grubs, 193 earth-worms, 25 caterpillars, and a mouse, — skin, bones, and all!

In the ocean depths we find that the marine worms, which constitute in the beauty of their appearance a magnificent assemblage, tunnel the upper crust of the bottom. Some years ago the moat or ditch surrounding Fort Jefferson, Fla., was pumped out, leaving a space nearly half a mile in extent, high and dry, which abounded in specimens that would have delighted the eyes of a specialist in any branch. Over this spot we had often, as a lad, enjoyed the venturesome fun of riding upon the backs of the great sea-turtles, kept there for the commissaries' use, had fished in every nook and corner, and now the opportunity was presented for penetrating below the surface of the bottom.

Some little digging showed, that, for a foot or more from the surface, the sand and mud was fairly alive with a variety of worms, numerous to an extraordinary extent, and in many cases beautiful beyond description. This condition of things is true, to a greater or less extent, in many localities; the worms retiring to the mud and other retreats during the day, at night venturing out, and even swimming at the surface.

If we take a drop of water from any ditch or pond, or even from the stem of a flower that has been standing in a vase, and place it under a microscope of even ordinary power, we shall find that it is a world of itself; a vast ocean, in fact, to the many forms that live there. Chief among these drop inhabitants, we notice numbers of little creatures that attract attention immediately. They resemble tall hats without brims, or crystal bags with fringed edges. And that they are busy bodies is at once evident, as they swim along at a won· derful rate of speed, eating as they go, keeping their fringes or *cilia*, which appear like so many arms, in perpetual motion; now bumping against each other, forcing their way among crowds of different animals, and always appearing full of life and energy.

These little creatures, invisible to the naked eye, are minute worms, or Rotifers; and among them we find some interesting light-givers. The *Synachata* is one; and others described by Ehrenberg, the largest being about one-eighth of a line in size, present a striking appearance under the glass in a dark room, — the little bags, seemingly at a white heat, darting about in every direction.

As small as are these wonderful creatures, they are well worthy of study; and even those not interested in natural history will find that the stems of their flowers, or the water in the vase, contain more wonders than they had dreamed of, — a single drop that can be lifted upon a pin-head being sufficient for the purpose.

The little hat-like form, *Hydatina senta*, already referred to, is remarkable for the rapidity of its increase. The eggs are laid or deposited within a few hours of the time they are first seen within the transparent parent, and twelve hours later the young break from the shell and appear; so that in a comparatively few days the descendants of a single animal might possibly far exceed the population of the United States. The larger worms are with hardly any exception ornamented in some remarkable way, and in many the splendors of their decorations must be seen to be appreciated. The radiating coronets of *Serpulæ* [30] are of the most delicate and beautiful description, abounding in bands and markings of striking hues. *Pectinaria* has upon its head a pair of combs that might be burnished gold; while *Eunicedæ* and *Nereidæ* [31] have equally resplendent decorations.

These charms of color, and they are of great variety, are seen by day; but at night many of these creatures assume the gift of phosphorescence, adding to the long list of marine light-givers that have been previously referred to. In four other families [32] are found the most beautiful light-givers of the group. Assuming that we have a certain species of the first mentioned in the aquarium, we may prepare for an extraordinary display. It is now snugly coiled up under a stone, perhaps fast asleep, and giving no evidence of its wondrous gift. Now touch it with the narrow handle of the dip-net, and a seeming electric spark is given out. But there is no electricity here : the light is a phosphorescent protest, and rapidly passes from scale to scale, until the whole animal stands out like a vivid shield of light against the bottom, glowing with the mysterious flame.

If the worm is greatly disturbed, we are presented with a unique method of protection. Upon feeling the blow or attack, the light becomes intense, and flashes quickly from segment to segment, and along all the series of *elytra ;* and, as the animal darts away, one or more of the scales become disconnected and are left behind, a luminous spot, to attract the attention of a possible follower, while the worm itself escapes.

Nearly all the phosphorescent worms are rapid swimmers, and noted for their agile movements ; and, as their scales are very readily disconnected, we may imagine in some cases a worm darting off and leaving a shower of sparks behind. In these worms the light is usually green.

We have seen that one of the deep-sea Crustaceans has phosphorescent bands upon its feet; and in the *Syllidæ,* a family which contains some remarkable worms, we find that the luminosity is confined to the under surface of the feet. In *Chætopterus* [33] a bright flashing light is emitted from the posterior feet, while a far more brilliant one glows at a point on the dorsum between the lateral wings of the tenth segment. The mucus of the animals appears to be the seat of the luminosity, and not only encircles the worm with an aureola of phosphorescence, but pervades the surrounding water with a rich bluish purple light, so vivid and brilliant

SEA PEN.
(*Pennatula.*)

BURROW OF PHOLAS.

that the medium in which the light-giver lives seems to have ignited, and to be slowly consuming its dependents.

It has been noticed, according to W. C. McIntosh, that an odor accompanies this display, resembling somewhat that pro·duced by phosphorus in combustion. We have noticed that many worms have a peculiar odor when handled, though not quite of this character.

The most brilliant of all these light-givers is *Polycirrus*, which emits over its entire surface a vivid pale-bluish light, marking it as one of the most beautiful of its kind; while *Sagitta* and many more add to the wonders in this generally considered uninteresting group of animals.[34]

CHAPTER VI.

LAMP SHELLS.

IN all the forms previously mentioned, the phosphorescence is conspicuous; but in the little bivalve Pholas it is almost hidden. The shells of the family *Pholadidæ* are noted for their boring habits ; penetrating into the hardest stone, as granite and gneiss, literally entombing themselves, as shown in Plate VII., Fig. 1, which represents a section of a block of granite into which the little animals have penetrated. How they can perform such a work, is something of a mystery ; but the foot, which is provided with a hard dermal protection, is probably the instrument used by the miner.

The most remarkable evidence of their work, according to Figuier, — though it is fair to say he has been disputed, — is seen in the Temple of Serapis on the Pozzuolan coast, where the pillars are perforated with holes, which this author claims were made by the Pholas,[35] when by a sinking of the crust the pillars were under water; the columns, by a reverse motion, having now re-appeared from the sea, bearing the evidences of their submersion.

As if to still further carry out the idea of the miner, the animal bears its own light, which, though vivid, could but little more than illumine the stony prison into which the Pholas has willingly ensconced itself. In Borneo, a fresh-water form has been found boring in the dead trunks of trees. Pliny was probably among the first to place on record the luminosity of this little borer, having stated that it shone in the mouths of those who ate it; and its phosphorescence has been studied by Réaumer, Beccaria, Marsilius, Galeatus, Montius, and others in modern times. One of Beccaria's experiments was to ascertain how the light affected certain colors. He secured a Pholas in a dark spot, and placed in its light ribbons of various colors. The white ribbon shone most brilliantly, the yellow next, and the green next, while others were so indistinct as to be hardly noticeable. Substituting liquids for the ribbons, the result was the same.

Beccaria also made one of the first practical applications of the phosphorescent Pholas, demonstrating that it could be used as a lamp. This was accomplished by placing one in seven ounces of milk, which rendered the latter so luminous that print could have been read by it, the milk appearing almost transparent. So it is within the bounds of possibility to write a *post-mortem* description of the Pholas by its own light.

It is evident from these simple experiments that the discovery of the secret of phosphorescence, and its practical application to the wants of mankind, would result in revolutionizing present systems, — a heatless, inexpensive, unextinguishable light being the perfection of possibilities in this direction, — and it is not improbable that the experimentalists of olden times may have had this in view when making their investigations. Both Réaumur and Beccaria attempted to render the light of this animal lamp permanent. By placing one in honey, the luminosity was apparently preserved for a year, the light re-appearing whenever the mollusk was placed

in warm water. Brandy extinguishes the light, and Galeatus
and Montius found that vinegar and wine produced the same
result. If the body of Pholas is heated slowly, the light
gradually becomes more and more intense, until, finally, at
45° Réaumur, or 56° Centigrade, it disappears, and cannot
be restored.

The secure position of the Pholas in its impregnable
fortress would hardly seem to require a warning or attractive
light; and its use must remain a mystery, though theory
could, of course, suggest explanations.

While the Pholas conceals its luminosity in its dungeon,
there are other molluscan light-givers which float about like
light-ships astray. These are Pteropods, or wing-footed mollusks; delicate fairy ships of marvellous beauty. By some
authorities they are said to represent the higher forms of the
Cephalophora, while others consider them as degenerate or
backsliding Cephalopods, of which the squids and octopi are
representatives. They are pelagic, free-swimming mollusks,
in which portions of the foot are modified into seeming
wings, so that the little creature seems to fly through the
water. They differ much in appearance. Some secrete a
glassy, horny, cartilaginous or limey shell, which in some
cases is only present in the larval forms, disappearing in the
adult; while others, again, preserve it through their entire
lives. The body is of various shapes : it is protected by the
shell when present, and can be drawn into it.

Though simple, helpless creatures, many have an arma-
ment which in a larger animal would be considered extremely effective. Thus in *Clio* each tentacle bears nearly
three thousand cylinders, each containing stalked suckers;
and, as there are six tentacles, the little animal can grasp its

PLATE VI.

PRAYA.

CLEODORA.

APOLEMIA.

microscopic prey with three hundred and sixty thousand hands. Besides this, it has a pair of many-toothed jaws and a toothed tongue. While extremely small, these animals exist in such vast multitudes, that they probably constitute an important food for certain whales.

One of the most interesting of the Pteropods, or wing-footed animals, as associated with our present subject, is the *Cleodora lanceolata* (Plate VI., Fig. 1). It has a pyramidal shell, terminating in three sharp spines, the wing-like fins rising above. It is rarely over half an inch in length, almost transparent, and bears in its shell a small light, which, however, is distinctly seen through the transparent covering. A more beautiful living lamp it would be difficult to imagine; and when slowly flying through the ocean, in countless myriads, they must present a wondrous sight. One of this genus, observed by Giglioli, emitted a very livid red light; the luminous organ being at the summit of the shell. There are many different genera and species. *Hyalea*, an oceanic wing-foot, moves very rapidly, and looks not unlike a butterfly darting here and there, in erratic flight, in search of food; but the little *Cleodora* moves in a regular and stately manner. In *Hyalea* observed by Giglioli in the harbor of Anjer, Java, the light, which contributed largely to the general phosphorescence, was confined to the basal part of the shell.

My young readers interested in geology are probably familiar with the curous *Conularia*, or cone in cone, which has been found in Australia sixteen inches in length, and has always been regarded a puzzle. It has been suggested that this is a gigantic fossil Pteropod. The little needle-like Tentaculites, from the Silurian and Devonian rocks, are also allies.

Some of the most remarkable mollusks are found among the sea-slugs, so called from their resemblance to the slugs of the garden. I have found them on the weed floating in the Gulf Stream, so resembling the latter in almost every particular that it was difficult to determine that they were not a part of the weed itself. *Scyllaea pelagica* is such a form; helpless, yet finding protection in its mimicry of the surroundings. Equally as remarkable is *Dendronotus*,[36] the bushy sea-slug whose gills resemble the branches of weed in a remarkable manner. This curious sea creature is quite common on the seaweeds of our New-England beaches. In the Mediterranean and Pacific is found the most unique of the group, the *Phyllirhoë bucephala* (Plate V., Fig. 2), which differs from many so entirely that it would seem to belong elsewhere. Like the other forms, it is pelagic, often being seen swimming along, resembling a fish, with its compressed body, and vertical, fan-like tail, and with long feelers or tentacles ahead. It is transparent and shelless in the adult stage, possesses no foot or *branchiæ*, evidently breathing through the body-walls or general surface. To add to its curious features, the *Phyllirhoë* is brilliantly phosphorescent; light being emitted from certain spots, shown in the engraving, rendering the tissues transparent and luminous. Examination has shown that the light proceeds from certain globular nucleated cells, which appear to be the terminations of nerves.

The *Phyllirhoë* thrives well in the aquarium, and has been studied and observed in the famous aquarium at Naples. When it is touched or is swimming, the light seems to diffuse the entire surface, so that it presents a striking contrast against the dark water; and undoubtedly this gift is a fatal one, attracting the attention of many a fish to the dainty morsel seemingly outlined in fire.

Giglioli refers to the luminosity of an undescribed Heteropod, the axis of whose body gave out a reddish light whenever the animal was excited. According to C. W. Peach, the young of *Eolis* are phosphorescent. Such instances where the animal is particularly defenceless are amusing refutations of the theories of naturalists who see in the light a warning.

The common garden slugs, the cousins of the snails, are well-known forms. They generally remain concealed during the day, coming out at night, and often doing much damage to vegetation which is largely laid to birds. I have kept many of them, and they offered an extremely interesting study. They secrete a remarkable amount of mucus, which they use in descending from a tree, just as a spider does its silk thread. The mucus exudes from the foot, passes along to the tail, when it is attached to the twig. This accomplished, the slug boldly launches itself into space, the thread becoming more and more attenuated, until finally, when the slug is near the ground, it is exceedingly fine. Nearly all our common slugs descend from trees in this manner, — quite a contrast to the slow, tedious ascent.

The amount of mucus that can be taken from them is remarkable ; and that it is also protective will be evident to any one who may experiment with them.

One genus, *Phosphorax*, found at Cape Verde, and, according to Duncan, at Teneriffe, has a luminous pore on the posterior border of the mantle. One species only is known, *P. noctilucus;* and its light has not, that I am aware, been made the object of any extended investigation.

The highest forms of the *Mollusca*, the Cephalopods, cuttle-fishes, are probably at times luminous. I have noticed what I presumed was a delicate, sensitive, luminous glow about an *Octopus* in a semi-darkened tank, but I am not satisfied to make the statement as fact. These forms are so remarkable for the waves of color that pass over them, and which seem to make them transparent, that one could readily be deceived.

The little *Cranchia* (Plate IV., Fig. 2) is a light-giver, its phosphorescence having been distinctly observed. It is an ally of the giant squids, which have been found fifty-five feet in length, and which, if luminous like their pygmy relative, would present a marvellous spectacle, darting veritable living arrows through the depths of the sea.

Giglioli refers to the phosphorescence of *Loligo saggitatus*, and to that of several small Octopods observed by him at Callao and Valparaiso. Their bodies gave out a pale whitish light, uniformly distributed.

LUMINOUS BEETLES, ETC.

a.—Lampyris splendidula—male.
b.— " " —female.

d.—Lampyris noctiluca—male.
e.— " " —female.

CHAPTER VII.

LIGHTNING-BUGS.

GEN. COUNT DEJEAN, aide-de-camp to Napoleon, was a most enthusiastic collector of beetles; and it is even said of him that he would march his army out of its way to pass through a good collecting locality. At all times during the campaigns which he helped to render famous, his attention was not taken from his favorite occupation; and his military cap was invariably conspicuous from the gorgeous beetles that were there immolated. Every one in the army, from the emperor down to his men, was aware of what was termed his weakness; and the latter were constantly on the lookout for specimens for their commander. At the battle of Wagram, 1809, the general went into the combat with his hat as usual ornamented with beetles, which he had received that morning; and, while standing near the emperor, a shot from the enemy struck him upon the head, knocking him senseless, and destroying his collection, — the hat being completely torn in pieces. The emperor, thinking him fatally wounded, hastened to his side, asking if he was still alive; upon which the general gasped out, "I am not dead; but, alas, my insects are all gone!"

The beetles are among the most interesting of all insects; and a study of them, though casual, will well repay my young readers, who cannot fail to be interested in their peculiarities, their habits, methods of protection and defence, their intelligence in caring for their young, and the wondrous light-emitting power of some species.

In my walks about the San Gabriel Valley, I generally meet a peculiar beetle, — a large, black fellow, who lumbers along in a clumsy manner. If touched, he cannonades me with a fluid of iodine color, which has a most disagreeable odor; so much so, that upon one occasion, my nostrils being in range, I was made temporarily faint by it. The fluid stained my hands like iodine, and caused not a little irritation to the skin. The beetle, then, is a living cannon; the fluid, which is contained in certain glands, being its defence. It can be ejected or thrown two inches, so that it affords quite a protection, and probably would be effective with birds.

Many insects have a curious odor which serves several purposes, — one, in rendering them nauseous to birds and various enemies; and, again, as a means of communication among themselves. Thus, if a community of deaf and dumb persons should decide to identify themselves by certain odors, we would see a practical application of this. One family would carry musk, and be recognized some distance off by it; and so with other perfumes or odors. This is just how some beetles call each other; and in the one referred to both male and female possess the same odor.

Some of the flesh-eating beetles (Plate IX.) exhibit great ingenuity and intelligence in securing a food-supply and an asylum for their young at the same time. To their work is due the fact that the remains of few animals are found at the surface. The moment the latter die, these insects, and especially the grave-diggers (*Necrophorus*), appear. They run about the body, if upon the ground, inspecting it with great interest. If the animal is small, and the earth about it not suitable for its purpose, it is removed to softer ground; and here the beetles begin to dig, undermining the body, until in a very few hours it has disappeared or been completely buried. I have seen a garter snake covered in four hours, and some animals are sunk in this way a foot from the surface. The beetles then feed upon the body, and the female deposits her eggs there, — perhaps thirty white cylindrical objects, which in time hatch; the young being in this way provided with an ample supply of food.

The Egyptian *Scarabæus*, noted for being found in the ancient tombs and monuments, and considered sacred by some of the natives, has an interesting method of caring for its future young. It encloses the eggs in round balls of various material suitable for food; a well is then dug several inches deep, into which the beetles roll the balls, then covering them: so that, when the young appears, it is encased in the food necessary to its existence.

Passing the giant beetles of the tropics, and many others that have features of interest, we come to the forms called lightning-bugs, which, of all their tribe, impress us as marvellous, and which are especially associated with our present subject.

> " Sorrowing we beheld
> The night come on ; but soon did night display
> More wonders than it veiled : innumerous tribes
> From the wood-cover swarmed, and darkness made
> Their beauties visible ; one while they streamed
> A bright blue radiance upon flowers which closed
> Their gorgeous colors from the eye of day;

Now motionless and dark, eluded search,
Self-shrouded; aud anon, starring the sky,
Rose like a shower òf fire."

Southey's description of the South-American fireflies does
not ill apply to the midsummer night festivals held in our own
woods and fields of the North, by the diamonds of the night.
As twilight deepens, these living lights appear; creeping
from beneath the bark of trees, out of the ground, or drop-
ping from some distant limb; darting here and there in
streams of light, soaring high in air, twinkling among the
leaves; while down in the hollow, where the cat-tails rustle
and nod, rises a veritable luminous cloud.

The producers of these displays are the lightning-bugs, —
beetles belonging to the family *Lampyridœ* (Plate X.,
Fig. 7). They are mainly of small size and soft texture;
the *larvœ* being flat and dark colored, and often presenting
the appearance of a bit of velvet. They are carniverous in
their habits, and can be found under stones and the bark
of trees. The velvet-hued *larvœ* of one species is often
seen on the surface of the snow, giving rise to stories of
worm showers. The family is divided, generally, into three
sub-divisions; and one, the *Lampyrinœ*, is noted for the
phosphorescence of many of the species. Numerous species
are known throughout the world and in this country, differing much in size; those in Kentucky and other Southern
States being somewhat larger than their Northern cousins.
In the South and the West-India Islands they are seen to best
advantage. In these isles of summer, especially Jamaica,
Gosse studied their habits, and observed their nocturnal
glories; and to him I am indebted for the following notes

relating to the West-India species. He says at all times their sparks, of various degrees of intensity, according to the size of the species, are to be seen, fitfully gleaming by scores about the margins of woods, and in open and cultivated places. He observed about fourteen species, all luminous. *Photuris versicolor*, a large species with drab-colored *elytra*, he found abroad soon after his arrival in December. One flying around the house in the evening, he was struck with its swift and headlong flight and nearly permanent luminosity, which was much more brilliant than that of any species he had at that time seen. The large *Pygolampis*, which he called afterwards *P. xanthophotis*, he did not observe until May, when one flew into his house at Bluefields one evening; and a few nights later he found them in great numbers on the very sea-beach at Sabito. It was conspicuous for the intensity of its light, much exceeding that of *Photuris versicolor*. Sometimes it is only the last segment but two that shows luminosity ; but, when excited, the whole hinder part of the abdomen is lighted up with a dazzling glare.

In June, in the woods of St. Elizabeth's, Gosse had special opportunities for observing the *Lampyridæ;* particularly along the road leading up the mountain from Shrewsbury to Content, where it is cut through the forest, which over-hangs it on each side, making it sombre even by day, and casting an impenetrable gloom over the scene by night. The darkness here, however, and especially at one point, — a little dell, which is most obscure, — is studded thick with fireflies of various species, among which the two large ones above named are conspicuous. *Pygolampis xanthophotis* he observed only in flight. Its light is of a rich orange color when seen abroad, but when viewed in the light of a candle appears yellow. It is not so deeply tinted as the abdominal light of *Pyrophorus noctilucus*, and is intermittent.

Photuris versicolor is noticeable by its frequent resting on a twig or leaf in the woods, when it will gradually increase the intensity of its light till it glows like a torch; then it gradually fades to a spark, and becomes quite extinct. It thus remains unseen for some time; but in about a minute, or it may be two, it will begin to appear, and gradually increase to its former blaze; then fade again, — strongly reminding the beholder of the revolving light at sea. The light of this species is of a brilliant green hue. Gosse says he has seen a passing *Pyg. xanthophotis*, attracted by the glow of a stationary *Phot. versicolor*, fly upward and play around it; when the intermingling of the green and orange rays had a charming effect.

The smaller species have, some a yellow, and some a green, light. *Pyg. xanthophotis*, when held in the fingers, will frequently illuminate a segment of the abdomen, over which the light plays fitfully, sometimes momentarily clouded, more or less, but generally saturated, as it were, with most brilliant effulgence. This species occasionally comes in at open windows at night, but much more rarely than the *Photuris versicolor* and the smaller kinds, a dozen or more of which may be seen almost every night, crawling up the walls, or flitting around the room and beneath the ceiling, of these Jamaica homes.

One of our commonest forms in the eastern United States is *Photuris pennsylvanicus*. It is about one-half of an inch in length, has a general yellowish color, with a few stripes or lines of brown or black. Both sexes have wings and quite long *elytra*.

PLATE X.

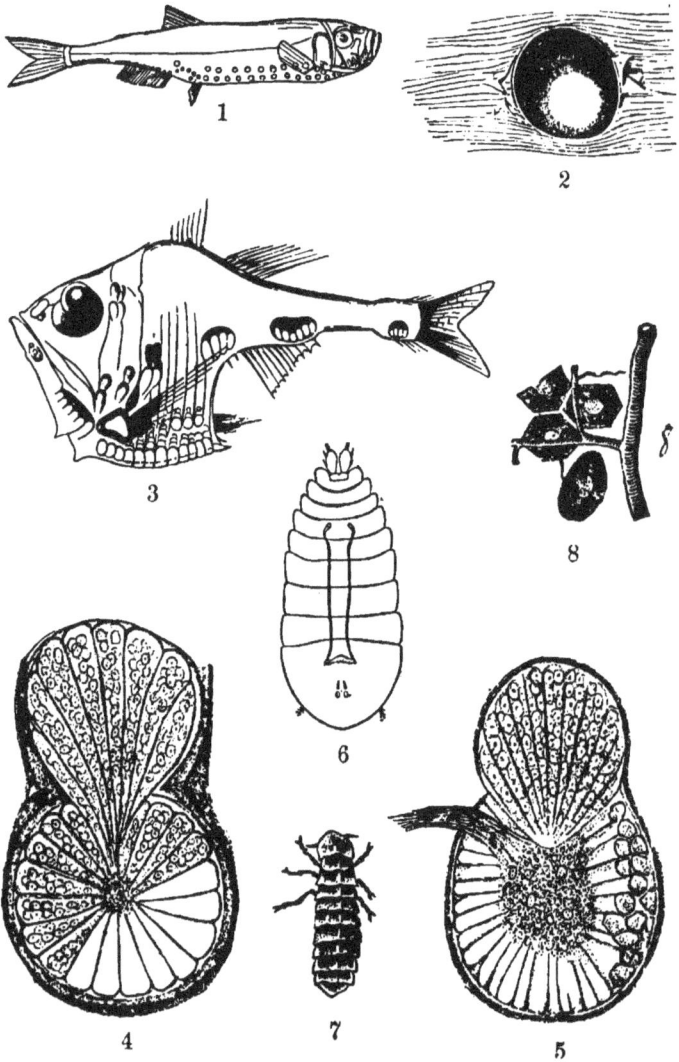

1. *Scopelus humboldti*, showing luminous spots.
2. Mother o'pearl organ from side of same.
3. *Argyropelecus*.
4. Longitudinal section of organs from abdominal region of same.
5. Luminous organ from nasal region of *Ichthyococcus*.
6. Luminous Crustacean. 7. Lampyris. 8. Light cells of same, and trachea (magnified).

In the diurnal *Lucidota*, often seen flying in shady places, and to be remembered by the peculiar, disagreeable, milky fluid they exude when caught, the luminous organs are feebly developed. In the female they are indicated by yellow spots found on the last ventral segment, and on the last two in the male. In the genus *Pyropyga* the light organs are inconspicuous, except in one species, — *luteicollis*. In *Pyractomena*, an attractive genus, this peculiar feature is well developed in both sexes, and the light vivid at times. The phosphorescent organs are larger in the male, and situated on the fifth and sixth ventral segments. Close examination will show in the male a large, stigma-like pore on each side, midway between the middle and the side, whose office is not perfectly understood. In the female the lanterns are at the sides of the segments. *P. lucifera*, found from Massachusetts to Texas, has extremely small luminous organs.

In the genus *Photinus*, certain species of which have parts of a roseate tint, the light-emitting organs are larger in the male than in the female, and vary considerably in position in the different species. In the male they cover the entire ventral segments, from the fourth to fifth inclusive; and on the fifth and sixth segments the little impressions or pores referred to are seen in the females. The light-organs occupy the middle portion of the ventral segments, and resemble a flat elevation upon the fifth segment. There are so many exceptions and differences, that the young naturalist will find it a particularly interesting study. Thus in *P. dimissus* the male has the usual illuminating apparatus, while it is entirely wanting in the female.

In the group *Lampyres* the lights are bright in the females, but variable in the males. For a long time only the male of

the genus *Phengodes* was known, the female being described
as another insect. The mistake was made owing to the fact
that the female never attains a development beyond the
larval condition, and is the only instance among beetles
where the larval female produces fertile eggs. The female is
about two inches in length, of a creamy-white hue in the daytime; but at night it presents a truly magnificent appearance,
emitting from the sides or margins of the segments a rich
green phosphorescent light.

Another light-giver rarely seen is the *larva* of *Mastinocerus*,
a slender, cylindrical form of a pale color. It lives upon
snails, and is feebly luminous. Mrs. King thus writes to Dr.
le Conte concerning it: " June 4, saw running rapidly over
the table, near a lighted lamp, a small Coleopter ; it was twisting its abdomen up over its wings, and evidently trying to
straighten them out, as they seemed moist and twisted at
their ends. The general appearance suggested *Mastinocerus ;*
and, acting on this thought, I captured it, and sat up till
a late hour to be assured of the truth. The insect was in a
small phial, and moved quickly. It gave out light conspicu-
ously from the head, feebly from the anal end, and still more
so from about the base of the abdomen. The light seen in
the head, though visible in the dark as a round spot, yet,
when taken into a room obscurely lighted, was invisible from
above; but, when the insect was suddenly thrown upon its
back, a light no larger than a pin-point was seen just about
the junction of the head and prothorax."

The method of illumination in this group is intermittent,
the light appearing as repeated flashes : hence the term "lightning-bugs" in contrast to the steady gleam of the fire-flies or
Elaters. Mr. A. E. Eaton has counted the flashes in *Luciola*

lusitanica, and found that there were thirty-six in a minute, each flash lasting from one-fourth to one-third of a second.[37]

The light of some species is intense, while that of others is very feeble. By placing detached parts of the luminous organs upon a page, I have been able to make out the type; and, if numbers of living lightning-bugs are confined, they can be utilized as a lamp, — rather a dull one, it must be confessed, unless the numbers are greatly augmented. The *larvæ*, as well as the *imagos*, are often luminous; even the eggs of some emit light.

An examination of the luminous organs during the day-time shows them to be yellowish or whitish patches on the various segments. If the hand is held over them, the light is seen, and in complete darkness they present a magnificent spectacle, — the light dying away, then growing intense, about the spot, so that it appears to be fairly trembling with heat, as if some chemical action was periodically asserting itself, causing the tissues to become suffused with a fiery glow; yet, if the most delicate thermometer is placed against the luminous organs of a large number of these insects, there is not the slightest elevation to show the presence of heat. If now we kill the insect, and remove the luminous matter, it resembles a bit of starch with luminous spots; and pressure, which admits more oxygen, causes a temporary increase in the light.

The luminous organs are similar in structure to the fat body of the insect, and are made up of light-emitting cells (Plate X., Fig. 8), surrounded by a maze of *tracheæ*, or air-tubes. In explanation of the light, it has been suggested that the cells secrete phosphuretted hydrogen, which becomes luminous upon contact with oxygen which reaches it through the minute air-tubes. Regarding *Luciola Italica*, Professor Emery says that the male *Luciolæ* gave out light in two distinct modes: in the night, when they are brisk and fly about, the light increases and decreases at short, regular intervals, so that it seems to twinkle. If one of them is caught flying, or disturbed in its rest by day, it shines less than at the maximum of its intensity when on the wing, but without intermission. It is remarked, however, that the luminous plates do not shine uniformly over their whole extent; but that sometimes one spot, and sometimes another, glows more strongly. If such a specimen is examined under the microscope, we perceive, on a dark background, bright, luminous rings, which are not, however, uniformly brilliant, but display certain more intense points, which flash up, and again disappear, or continue to shine on faintly for a time, re-appearing afterward in full splendor. These changes take place without any regular succession.[38]

The common lightning-bugs of Europe are *Lampyris noctiluca* (Plate IX., Fig. *d*) and *L. splendidula* (Fig. *a*). Their life history is an interesting study, and a brief description will apply to all. In early spring we find the little yellow eggs, perhaps gleaming with the wonderful phosphorescence, and thus finding protection, attached to blades of grass or other objects just above ground. The *larva* (Plate IX., Fig. *c*), a long, narrow, flat creature, soon appears and begins a predatory life; even being provided with an apparatus for removing the mucus of its victim. About the month of April it attains its full vigor, and during the summer changes to the pupa form, or hibernates all winter, entering a deep sleep, and assuming its new shape the following spring. We see the light from the very first in the eggs of some;

then in the *larva*, there appearing like little sacs on the under surface, one on each side of the middle line, so arranged that the insect can hide them by retracting the body, and causing them to blaze out when the abdomen is extended. Nothing in all nature is more wonderful than the changes through which these and other insects pass before attaining adult growth.

The *larva* is a busy little creature, full of life; but, when about to change, it becomes lethargic and quiet, as if impressed with the importance of the coming metamorphosis. Finally it wriggles out of its old skin, and becomes a *pupa*, also luminous; exceedingly lively, yet with its motions restricted. It moves its *antennæ* and legs, and pushes itself along by movements of the abdomen. Finally the perfect insect appears, with its wondrous array of lights, so little understood, and which, if accompanied with the ordinary amount of heat attendant upon such a display, would soon roast or fry its possessor. As to the use of the lights, we can only conjecture. It has been shown that one insect recognized the other by it, and thus it may be a sign language; while, according to others, it is a warning to birds and other enemies.

Mr. Darwin thus refers to the lightning-bug of South America: "All the fire-flies which I caught here (at Rio) belonged to the *Lampyridæ* (in which family the English glow-worm is included), and the greater number of specimens were of *Lampyris occidentalis*. I found that this insect emitted the most brilliant flashes when irritated; in the intervals, the abdominal rings were obscured. The flash was almost co-instantaneous in the two rings, but it was just perceptible first in the anterior one. The shining matter was fluid and very adhesive; little spots, where the skin had been torn, continued bright with a slight scintillation, whilst the uninjured parts were obscured. When the insect was decapitated, the rings remained uninterruptedly bright, but not so brilliant as before. Local irritation with a needle always increased the vividness of the light. The rings in one instance retained their luminous property nearly twenty-four hours after the death of the insect. From these facts it would appear probable that the animal has only the power of concealing or extinguishing the light for short intervals, and that at other times the display is voluntary. On the muddy and wet gravel walks, I found the *larvæ* of *Lampyris* in great numbers. They resembled in general form the female of the English glow-worm. These *larvæ* possessed but feeble luminous powers; and on the slightest touch they feigned death, and ceased to shine; nor did irritation excite any fresh display."

PLATE XI.

LUMINOUS BEETLE. *(Pyrophorus noctilucus.)*
In burrow of Mole Cricket.

CHAPTER VIII.

FIRE-FLIES.

SOME years ago an American gentleman, visiting in one of the large cities of South America, was invited to a masquerade ball at one of the finest private residences in the city. The ball-room was the garden, — a veritable fairy-land abounding in plants of the most novel and beautiful description, — and upon the grass had been laid an extended platform for the dancers. It was moonlight when the festivities began, and no artificial lights were used; yet at various intervals among the flowers soft gleams appeared, apparently for ornament. Among the first comers was a tall gentleman dressed in a style of several centuries ago, a most picturesque costume; but what particularly attracted the attention of the American were the decorations of this gentleman and his companion. Around the broad-brimmed hat he wore a band of what appeared, from a distance, to be gems, that flashed like diamonds, presenting a magnificent appearance. The lady's costume was still more remarkable, being fairly ablaze with these brilliant scintillations. As the evening wore on, he was presented to these maskers, when he found that the light proceeded from innumerable luminous insects which had been secured by delicate wires, and fastened upon the hat and the lady's dress.

About the garden, hundreds of the insects were confined in delicate glass globes, which without emitting much light, added to the charm and novelty of the surroundings.

In Vera Cruz these beetles are so commonly used as toilet ornaments that they form an important article of trade; and the natives make a business of catching them, and in a way that would seem to show that the lights of insects are their means of recognition. The fire-fly hunters provide themselves with long sticks, upon the end of which is fastened a burning coal. This waved in the air attracts the light-givers, and they are entrapped in a net. They are then placed in a box covered with a wire netting, bathed twice a day in tepid water, and at night fed with sugar-cane.

The insects utilized in this curious manner are fire-flies, — distinguished from the lightning-bugs by the steady glare they produce. And that the lights of these Elaters, as they are scientifically called, is intense, and of practical value in other ways, we may realize from the statement of Professor Jaeger, who says, " I feel particularly grateful to these little insects, because, during my excursions in St. Domingo, they were frequently the means of saving my life. Often has dark night surrounded me in the midst of a desert forest, or on the mountains, when the little animals were my only guide ; and by their welcome light I have discovered a path for my horse, which has led me safely on my journey." If a number are confined in a glass, they emit sufficient light to read by.

It is in the genus *Pyrophorus* that we find the most remarkable light-givers; the different species being found principally in tropical America. In Plate XI. *Pyrophorus noctilucus*, a form common in the West Indies and Brazil, is shown. It ranges from 1.50 to 1.75 inches in length; is a black or rusty-brown color; and, if observed during the day-light, two conspicuous oval spots of a yellowish white hue are seen on each side of the prothorax. These are the lan-terns of the Elaters, and in the dark glow with a brilliancy far exceeding that of the lightning-bugs. These lights shine from above, while between the part known as the metathorax and the first abdominal segment gleams another, or lower light, even more brilliant than the other: so the *Pyrophorus,* turn which way it will in its flight, emits a flash of light. The light appears to be dependent upon the will, as when feeding or asleep it is not seen ; attaining its greatest bril-liancy during activity and flight. The color of the light, as seen by the author, is a rich green ; but the eggs emit a light of a bluish tint, according to Dubois. This naturalist has made some extremely interesting experiments with this beetle. The eggs which he dried retained their luminosity for a week, the light re-appearing when they were placed in water. He ground the luminous organs in a mortar, after having dried them in vacuum, and then mixed them in boiled water ; the latter immediately becoming luminous. Dr. Du-bois concludes that the light of the *Pyrophorus* is intended as an illuminator for itself alone. To prove this, he covered one of the upper lights with wax, and the animal moved in a curve ; when both spots were covered, the beetle soon stopped, and then moved in an uncertain manner, carefully feeling the ground with its *antennæ.* The spectrum of the light was extremely beautiful, being continuous, without dark or brilliant rays; and, what appears most remarkable, the com-position of the light was found to change with its intensity. As to the exact cause of the light, how it is produced, the secret yet rests with Nature.

Dr. Kidder thus refers to the brilliancy of one of these wondrous light-givers: "Before retracing my steps, I stood for a few moments looking down into the Cimmerian blackness of the gulf before me; and, while thus gazing, a luminous mass seemed to start from the very centre. I watched it as it floated up, revealing in its slow flight the long leaves of the palm *Euterpe edulis*, and the minuter foliage of other trees. It came directly towards me, lighting up the gloom around with its three luminosities, which I could now distinctly see."

The insect was the *Pyrophorus noctilucus*; a longish click-beetle of a dull blackish-brown color, and covered over with a short, slight-brown pubescence. When walking or at rest, the chief light that it emits proceeds from the two yellow tubercles on the thorax, so conspicuous in dead specimens; but, when flying, another luminous spot is discernible on the hinder part of the thorax, and this is continued to the under side of the insect.

Ovideo says that the Indians travel in the night with these insects fixed to their hands and feet; and that they spin, weave, paint, dance, etc., by their light. In Prescott's "Conquest of Mexico," we are told that in 1520, when the Spaniards visited that country, "the air was filled with the *Cucujo*, — a species of large beetle, which emits an intense phosphoric light from its body, strong enough to enable one to read by it. These wandering flies, seen in the darkness of the night, were converted, by the excited imaginations of the besieged, into an army of matchlocks."

At the time of the discovery of Hispaniola, Peter Martyr assures us that the natives, in their night journeyings through the woods, were in the habit of fastening a number of these

PLATE XII.

THE LANTERN FLY.

(*Fulgora lanternaria.*)

According to Madame MERIAN, Marquis SPINOLA, and others.

light-givers to their feet to light the way. On this occurrence Southey founds the incident mentioned in " Madoc "
where Coatel guides Madoc through the cave: —

" She beckoned, and descended, and drew out
  From underneath her vest a cage, — or net
  It rather might be called, so fine the twigs
  Which knit it, — where, confined, two fire-flies gave their lustre."

CHAPTER IX.

LANTERN-FLIES.

WHEN Sir Charles Lyell visited this country some years ago, he expressed much interest in the sea-serpent question; and one of his first inquiries, when introduced to a certain gentleman, was, " Have you heard any thing about the sea-serpent? " The reply was, " Unfortunately I have seen it."

If Mme. Merian were alive, and a similar question should be propounded to her regarding the luminosity of the South-American lantern-fly, she could with propriety make a like response. She makes a definite and distinct statement concerning the phosphorescence of the *Fulgora lanternaria*, yet to-day it is declared non-luminous by nearly all scientists.

It is not our intention to champion the cause of this enthusiastic naturalist; but to some it would seem that the direct evidence of a single observer of good repute should have some weight against an indefinite number who merely failed to corroborate the observation. In the chapter on luminous plants, an almost similar instance is given, where for years the direct statement of the daughter of Linnæus regarding the luminosity of a plant was doubted by scientific men, until finally a well-known botanist confirmed it. To some it would seem possible that the *Fulgora* emits light only at certain times, and under peculiar conditions; be this as it may, scientific opinion 'is entirely against its luminosity, and the light in the figure of *Fulgora lanternaria* (Plate XII.) is introduced merely to show its supposed appearance according to the description of Mme. Merian and her supporters. The question is so interesting, and so typical of many that arise, that we introduce the opinions of the various authorities upon the subject.

The two most interesting species come from China and South America, — *Fulgora candelaria* from the former, and *F. lanternaria* from the latter. The Asiatic species is the smallest, measuring about two inches in length, and noticeable for the peculiar horn-like projection on the head, supposed to be the luminous organ. Its colors are rich and attractive; the head and proboscis, as we may call it, being a fine reddish brown, apparently dotted here and there with white specks. The thorax is a deep yellow hue; the body, black above, and yellow beneath. The wings are still more striking, — the upper pair dark, with many green reticulations, that divide the entire surface into many minute squares, yellow spots being scattered here and there; the under wings are orange with black tips.

The *Fulgora lanternaria* of South America is nearly three inches and a half in length from the tip of the head to the extremity of the tail, and about five inches and a half broad with its wings expanded. The body is of a lengthened oval shape, sub-cylindric, and divided into several rings or segments; while the head is distinguished by a singular prolongation, which sometimes equals the rest of the body in size. The general color is yellow, variegated with many brown stripes and spots. The wings are large and powerful; the lower pair ornamented with a large eye-spot, well shown in the accompanying figure; the iris or border being red, while the centre is half red and half white, rendering it a very conspicuous object. The remarkable extension of the head — or lantern, as it has been called — is pale yellow, ornamented with longitudinal red stripes. In this projection the luminous property of the lantern-fly is said to exist.

In Mme. Merian's work on the insects of Surinam, she says, "The Indians once brought me, before I knew that they shone at night, a number of these lantern-flies, which I shut up in a large wooden box. In the night they made such a noise that I awoke in a fright, and ordered a light to be brought, not knowing from whence the noise proceeded. As soon as we found that it came from the box, we opened it, but were still much more alarmed, and let it fall to the ground in a fright, *at seeing a flame of fire* come out of it; and as many animals as came out, so many flames of fire appeared. When we found this to be the case, we recovered from our fright, and again collected the insects, highly admiring their splendid appearance."

Such a statement naturally attracted attention; and, from its publication until the present, collectors have endeavored to substantiate it. Count Hoffmansegg states that his insect collector Sieber, who was a practical entomologist of thirty years' experience, took many specimens of *F. lanternaria* in Brazil, but never saw one emit light. The Marquis Spinola, in the annals of the "Entomological Society of France," vol. xiii., contends for the luminosity of the entire tribe. On the other hand, M. Richard succeeded in raising a species of *Fulgora*, but failed to observe the light; while M. Westmael assures us that a friend of his observed the luminosity.

John C. Branner, Ph.D., states that when in South America he was often informed that it was luminous, but never could find any one who had personally seen the light. Snr. Luiz A. A. de Carvalho, jun., of Rio de Janeiro, who had fine specimens in his cabinet, assured him that he knew of no evidence whatever that they produced light. In the article on phosphorescence in the last edition of the Encyclopædia Britannica, Mr. William E. Hoyle, F.R.S., of the "Challenger" expedition, apparently accepts the *Fulgora* as a light-giver; as he says, "Whilst the lantern-flies, *Fulgoridæ*, carry their light at the extremity of a long, curved proboscis." Professor P. Martin Duncan writes, "It is doubtful if the *Fulgora*, so often described in books as the lantern-fly, has a scarlet light, if any at all."

The *Fulgora* is not remarkable for its supposed light alone; as in Brazil, where it is called *Gitiranaboia*, etc., it is considered by some natives to be extremely deadly. Mr. John C. Branner, of the Indiana University, investigated the subject, and found that the natives believed that the long proboscis was the poisonous organ; and that when this struck any animal, no matter how large or powerful, the latter immediately dropped dead. Even a distinguished Brazilian engineer assured Mr. Branner of the truth of the stories, saying that monkeys were often seen to fall dead from trees along the Amazon, killed by the deadly *Gitiranaboia;* and a local paper reported the fact that these insects were destroying cattle in the southern provinces. At Parà, Mr. Branner was assured that a child died in great agony after being stung by one. It is needless to say that the lantern-fly is perfectly harmless, and its poisonous properties as mythical as modern science deems its light.

Concerning the Chinese and African species, there is the same conflict of opinion. Dr. Phipson, an eminent authority on phosphorescence, evidently accepts them as luminous; as, in referring to the proboscis, he says, " It is from these appendages, the sides of which are transparent, that the phosphoric light appears ; " and in mentioning *Fulgora cande-laria*, he says, without giving his authority, " It is said, also, that the trunk of a tree covered with numerous individuals of *F. candelaria*, some in movement, others in repose, presents a very grand spectacle, impossible to describe, but which may be witnessed sometimes in China." Dr. Donovan, in his "Insects of India," figures the *Fulgora pyrrhorynchus;* and Phipson states, " It is said to emit a light of a fine purple color. Donovan evidently had some reason for believing that they emitted light, as he represents them in the act.

In " Packard's Guide," there occurs the following reference to the light of an East-African *Fulgora:* " Mr. Caleb Cooke of Salem, who resided several years in Zanzibar, Africa, told me that the lantern-fly is said by the natives to be luminous. They state that the long snout lights up in the night, and in describing it say, its head is like a lamp (*keetchwa kand-tah*)." According to William Baird, Esq., there is an edict in China against young ladies keeping lantern-flies. Altogether, the question is quite in keeping with the mystery that surrounds the entire subject of animal phosphorescence.

One of the classes into which the insects are divided is termed *Myriopoda*, from the fact that the individuals which compose it are supplied with a seemingly endless number of locomotive organs. The centipedes and millepedes, "hundred" and "thousand legs," are the names by which they are most commonly known. The body is long and cylindrical in the genus *Geophilus*, being made up of from thirty to two hundred segments, each bearing a pair of short feet. In the Eastern States *Lithobius Americanus*, Wood., is perhaps the most familiar form, and often found under old logs.

Some of the centipedes are very poisonous. Such a one is *Scolopendra heros*, Girard., the poison being stored in two enormous fangs. In Southern California I have found extremely large specimens of this genus. In the East Indies *Scolopendra gigantea*, Linn., attains a length of nine inches, and is a most repulsive appearing creature, and so dreaded that the most extravagant stories are told as to its power. A native informed me, who evidently believed his statement, that a man died near him from having one merely walk over him. The bite is undoubtedly poisonous, as is that of many of our common spiders; but I never could find an authentic case where it had resulted fatally.

As hideous as they are in certain parts of South America, a huge species, which attains a length of a foot, is eaten; the native children, according to Humboldt, tearing off the heads, and devouring the remainder with evident enjoyment.

There are about eight hundred species of *Myriapods*, and among them is one, the *Geophilus electricus* of Europe, that is positively luminous; though Phipson, referring to them as *Scolopendræ*, gives two luminous species, *S. electrica*, Linn., of Europe, and *S. phosphorea* of Asia. Specimens of the former, observed in fields at night, have been compared to minute pieces of red-hot coal, so vivid was the light. Probably the finest spectacle of the luminosity of these insects was observed by M. Audouin, at Choissy-le-Roi, near Paris. Noticing a light upon the ground in a chiccory field, he ordered his man to turn up the earth, when the scene that followed is described as truly magnificent. The soil appeared as if it had been sprinkled with molten gold, the display being intensified if the insects were trodden upon or rubbed; in the latter case, streaks of light appeared, as if a bit of phosphorus had been placed upon the hands, the light being distinctly visible for twenty seconds.

The *Geophilus electricus* (Plate XIII., Fig. 2) is a small, inconspicuous insect, about an inch and a half in length, and one tenth of an inch in diameter. Like others of its kind, it lives in holes in the ground, and, when discovered, makes off rapidly by the use of its one hundred and forty legs. The interesting fact that the luminous secretion could be separated from the insect was originally noticed by Macartney seventy years ago, who found that the fluid, as he terms it, could be communicated by the centipede to every portion of its integument. This author also claims that the insect is only luminous after exposure to the sun, — a peculiarity that is found in certain minerals described in a later chapter. The most remarkable exhibition of the luminosity of these insects has been recorded by Mr. B. E. Brodhurst, who saw it first twenty paces away, so vivid was its display. The light looked like moonlight, so bright was it through the trees. " It was a dark night, warm and sultry. Taking a letter, I could read it. It resembled an electric light, and proceeded from two centipedes and their trails. The light illuminated the entire body of the animal, and seemed to increase its diameter three times. It flashed along both sides of the creature in sections, there being about six from head to tail between which the light played. The light behaved precisely like the electric light; moving, as it were, perpetually in two streams, one on each side, and yet lighting up the whole body. The trail extended one and a half feet from each centipede over the grass and gravel walk, and it had the appearance of illuminated mucus. On securing one of the creatures for examination, I found, on touching it, the light was instantly extinguished." This observer says that this phenomenon was frequently seen by others about his place.

Mr. Brodhurst continues, "Professor Flower identified the centipede as *Geophilus subterraneus*. The published descriptions of the luminous properties of British centipedes differ considerably from what I observed in this instance: the latter attributing light to the creatures only when disturbed. I was never able to induce my centipede to shine while in captivity."

CHAPTER X.

BY CRAB-LIGHT.

IN drifting over the calm waters of the ocean as night comes on, we notice in the depths below luminous forms of infinite variety. These are *medusæ*, as we have seen, moving here and there like veritable comets. They approach so near the unruffled surface, at times, as to expose the gleaming disk. The nets of the fishermen come up entangled in their golden trains, and along shore processions and columns of these wondrous shapes pass and repass.

As the night grows apace, and the wind rises, they sink into the deeper waters: yet the foam and crest of the waves still give out the curious light, though now from another source. Much of this is due to Crustaceans, minute creatures often almost invisible to the naked eye, yet possessing this wonderful gift of phosphorescence to a marked degree.

Some species of the little *Gammarus* are remarkable for their clear silvery light. They are familiarly known as water-fleas, attracting attention from their leaping powers, and are often found under seaweed above high-water mark, darting here and there in incredible numbers when their home is disturbed. These forms are extremely valuable as scavengers.

That these interesting animals were light-givers, has long

been known; Viviani observing it in a number of species in the beginning of the present century.[39] There is one peculiarity about many of these small animals; that is, the light has a more decided red tint than that of any other group of animals. This is especially true of many of the water-fleas, or *Entomostracans*, and the extremely transparent, ten-footed kinds. The light is often intense, but fitful and shortlived. It seems to start from the locality where the legs join the body, and rapidly spreads beneath the skin until the entire body appears to be suffused with light, and the little animal consumed with an internal fire. Yet if a bushelful of these gleaming living lights were confined, and a thermometer placed among them, it would not show the slightest variation or evidence of heat. The little Cyclops is very common in our fresh-water ponds, and forms a beautiful object under the microscope.

Along our sea-shores we may often see, under the rocks, clinging to the eel-grass, or among the thickly growing stems of *Coralina officinalis*, in some pool left by the tide, gleaming spots that move about in an erratic manner; now many collecting together, then breaking up into small patches of light, which in turn separate again. They are curious Crustaceans, known scientifically as the *Idotea phosphorea*. By day we shall find that they are usually spotted or entirely a bright yellow; at night emitting fitful gleams, perhaps as signals or as means of communication to their fellows.

In the Arctic regions beautiful lights have often been seen, due to a minute crustacean. Lieut. Bellot first observed it in the North-American polar regions, and Nordenskiöld refers to it in his "Voyage of the Vega." The most brilliant displays have been seen at Mussel Bay. Nordenskiöld says,

" If during winter one walks along the beach on the snow, which at ebb is dry, but at flood-tide is more or less drenched through with sea-water, there rises at every step an exceedingly intense beautiful bluish-white flash of light, which in the spectroscope gives a one colored labrador-blue spectrum. This beautiful flash of light arises from the snow, that shows no luminosity before it is stepped upon. The flash lasts only a few moments, but is so intense that it appears as if a sea of fire would open at every step a man takes. It produces, indeed, a peculiar impression on dark and stormy winter days. The temperature of the air is sometimes in the neighborhood of freezing of mercury. It is certainly a strange experience to walk along in this mixture of snow and flame, which · at every step one takes splashes about in all directions, shining with a light so intense that one is ready to fear that his shoes or clothes will take fire. If carefully examined, the cause of this phenomenon is found to be a little crustacean, *Metridea armata*, that somewhat resembles the Cyclops. The great changes of temperature to which it is subjected in the snow-sludge seem not to affect it."

Few phosphorescent animals exhibit their glories during the day; but *Sapphirina* (Plate X., Fig. 6) is an exception. It is one of the largest of the *Entomostracans*, about a quarter of an inch in length, broad and flat, without the beauty of form which characterizes Cyclops, Calanus, and others; but what it lacks in this respect is more than compensated by its marvellous powers of light production, few animals of any kind equalling it. So vivid is the phosphorescence, that it can be distinctly seen by day; and, peering down into the depths where it abounds, flashes of color — blue, gold, sap-

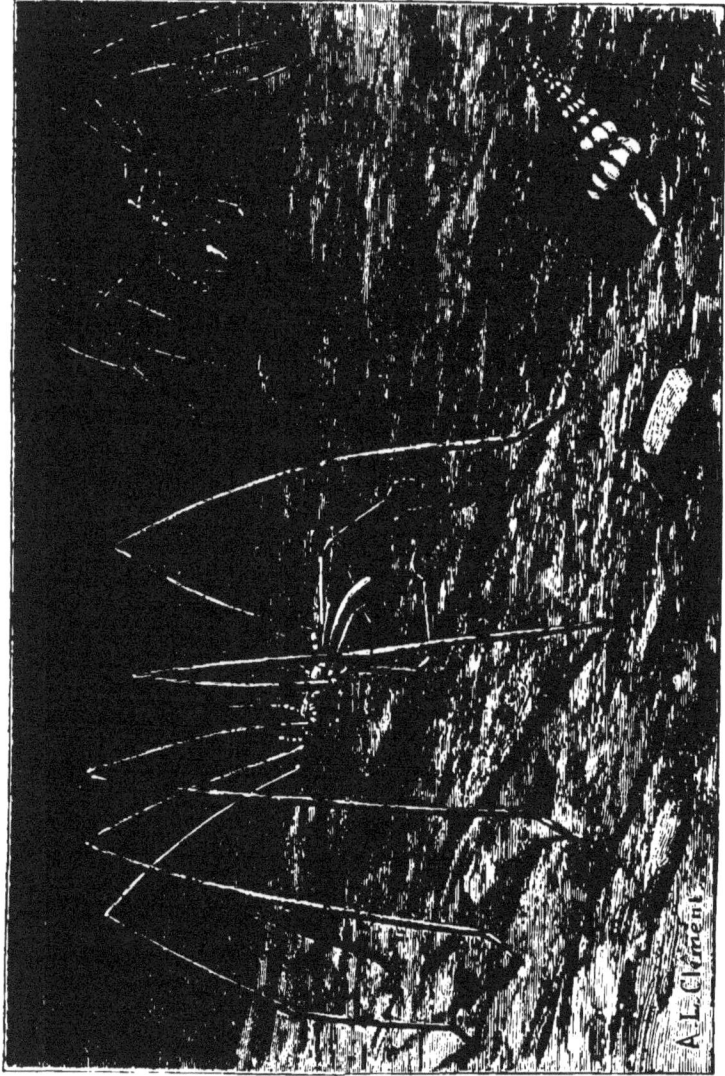

SPIDER CRAB.
(*Colossendeis.*)

A. L. Clement

phire, purple, green, and other hues — appear in bewildering frequency, ranging from the softest to the most intense and vivid lights, marking this living sapphire as one of the true gems of the sea.

Giglioli mentions an Isopod crab, brilliant with gold and purple, gorgeous with iridescence, and possessed also of the additional charm of phosphorescence. The light-emitting organs in the *Entomostracans* observed by him were in the anterior portion of the thorax.

The young (Zoëa) of the graceful little opossum shrimp *Mysis stenolepis* is phosphorescent. The adult forms are extremely interesting objects for study, the eggs and young being carried in a little pouch beneath the thorax. Allied to this little sea-opossum is *Lucifer*, that is to the crustaceans what the walking-stick is to the insect world; a veritable incongruity, resembling a branch of weed, and doubtless finding some protection in the mimicry. Some specimens, according to Giglioli, are luminous; the gift perhaps forming a signal language, a code understood in this world under the sea. The position, or seat, of the luminosity in crustaceans differs as widely as the intensity and color of the light; and in the little *Stomatopod*, formerly considered as an adult, and described as *Squillerichthus*, we find the culmination of wonders, as, in a specimen of this genus found in the Atlantic, the seat of the brilliant intermittent yellowish-green light is in the eye-stalk; so that the eyes themselves may be said to be veritable lanterns.

The phosphorescence of crabs was probably observed for the first time by Sir Joseph Banks, on his voyage from Madeira to Rio Janeiro; a small crab, named *Cancer fulgens*, being captured, which was remarkably luminous. Sir Joseph does not state whether the light came from the entire body or was confined to certain localities. MM. Eydoux and Souleyet, naturalists of the French exploring ship " La Bonite," noticed a small luminous crustacean, and succeeded in separating the phosphorescent secretion from the animal. They describe it as yellowish, viscous, and soluble in water, and found that its luminous properties soon disappeared. It was their opinion that certain crustaceans secreted the luminous matter, and that they differed much in their method of producing it. Certain small crabs, they believed, could display a certain amount of light when irritated; the phosphorescence at these times appearing in jets, forming a cloud or halo of light in which the animal seems to disappear.

In the abyssal depths of the ocean, where probably no ray of sunlight reaches, the crabs are possibly all luminous. Many of these deep-sea forms have a wide geographical distribution. Thus the Lithodes are found from the shallow waters of the north and south poles to the tropics, in the latter living in a region over which rests three-quarters of a mile of water. Many other crustaceans live in depths vastly more inaccessible than this, and under a much greater pressure. Thus _Colossendeis titan_, a strange creature, whose stomach is prolonged to the ends of the feet, is found living at a depth of about two miles and a half. These creatures, a species of which is shown in Plate XIV., are the spiders of the sea, resembling their not distant allies of the land, at least in appearance.

The different depths affect the inhabitants to a more or less extent. In some, the eyes seem to have lost their proper functions; and an instance is thus described by the Rev.

A. M. Norman, naturalist of the "Porcupine," the crustacean being *Ethusa granulata:* "The examples at one hundred and ten to three hundred and seventy fathoms in the more southern habitat have the carapace furnished in front with a spinose rostrum of considerable length. The animal is apparently blind, but has two remarkable spiny eye-stalks, with a smooth rounded termination where the eye itself is ordinarily situated. In the specimens, however, from the north, which live in five hundred and forty-two and seven hundred and five fathoms, the eye-stalks are no longer movable. They have become firmly fixed in their sockets, and their character is quite changed. They are of much larger size, approach nearer to each other at their base; and, instead of being rounded at their apices, they terminate in a strong rostrate point. No longer used as eyes, they now assume the functions of a rostrum; while the true rostrum, so conspicuous in the southern specimens, has, marvellous to state, become absorbed. Had there been only a single example of this form procured, we should at once have concluded that we had found a monstrosity; but there is no room for such an hypothesis by which to escape from this most strange instance of modification of structure under altered conditions of life. Three specimens were procured, on two different occasions, and they are in all respects similar."

Specimens of these crabs found in *shallow* water had perfect eyes; but, beyond one hundred and ten fathoms, they had changed as above stated. As Darwin has said, the stand for the telescope is there, though the telescope with its glasses has been lost.

Probably many of the deep-sea forms are luminous in some way.[40] *Aristeus* and allied forms are known to have phosphorescent eyes. Others have phosphorescent organs in various parts of the body. In one, the legs bear luminous bands that sparkle and gleam as the animal moves along in its dismal home. In others there are certain globular luminous organs beneath the thorax, and between the abdominal swimmerets that have been described as eyes. The light emitted by the several organs is of different degrees of brilliancy.

Vaughn Thompson is opposed to the theory that the objects on the side of the trunk, and along the ventral face of the tail, of these little creatures are eyes. " A re-examination," he says, "proves that they are not visual organs at all, but constitute rather a highly complicated luminous apparatus together; the lenticular body of the organs acting as a condenser, which, in connection with the great mobility of the globules, enables the animal to produce at will a very bright flash of light in a given direction. The great majority of species possess these organs, generally arranged in a perfectly similar manner; but in a large, deep-sea, non-pellucid *Euphausia*, V. Willemoes Suhm could not detect these globules in their usual place.

" The phosphorescent light emitted by the species of the *Euphausiidæ* was frequently under observation. One taken by forceps exhibited a pair of bright, phosphorescent spots directly behind the eyes; two other pairs were on the trunk, and four other spots were situated along the median line of the tail, — all quite visible to the naked eye. The light of these is a bluish white. After a brilliant flash as been emitted from the organs, they glow for some time with a dull light. The light is given out at will by the animal, and usually, but not always, when irri-

LUMINOUS CRUSTACEANS.
Nematocarcinus gracilipes.
Cyclops (magnified).

tated. The most brilliant flashes occur when freshly taken
from the sea. Under the microscope these phosphorescent
organs appear as pale-red spots, with a central, clear, lenticular body. The light comes from the red pigment surrounding the lenticular space. Mr. Murray observed at night, on
the surface of the sea in the Faeroe Channel, large patches
and long streaks of apparently milky-white water. The tow-
nets caught in these immense numbers of *Nyctiphanes nove-
jica*, and the peculiar appearance of the water seemed to be
due to the diffused light emitted from the phosphorescent
organs of this species.

Many of the deep-sea shrimps are remarkable for their
brilliant coloring. *Aristes* is a bright red, with *antennæ* five
or six times as long as its body. Equally strange is the
long-legged *Nematocarcinus* (Plate XV., Fig. 1), and the *Oplo-
phori* and *Notostomi*, curious little creatures, that have no
common names, are of an intense red hue, while others are
brown, rose, or spotted with red; showing that Nature
decorates her own even in the uttermost depths of the sea.

CHAPTER XI.

SEAS OF FLAME.

IN the summer months in tropical and semi-tropical waters, often during several days in succession, the ocean presents a surface almost unruffled. The fin of some roving shark, the splash of the flying-fish, or, if near shore, the plunge of the pelican or gull, are the only objects that disturb the sea of glass. At such times, after the sun had gone down, we have lain in our boat, with faces as near the surface as possible, and watched the wondrous panorama of the submarine world. Here great globes of light seemed to shoot through the watery space: every fish left a train of light; while the dolphin, or other great forms, gliding by, appeared converted into fiery monsters; and, as they rose to the surface, fountains of phosphorescence burst from the sea.

The forms which tend to produce this remarkable appearance in the ocean depths are many; but, in the warm waters of the tropics, the most noticeable are those belonging to the class known scientifically as *Tunicata*. Aside from their luminous properties, the Tunicates are extremely interesting, from the fact that they are now supposed to represent, with perhaps one exception, the lowest form of backboned life, — being what are called degenerate forms. In the larval stage of some species a noto-cord is present, which is supposed to represent the backbone of higher vertebrates. In some, when the animals assume the adult form, the little spinal cord is absorbed; but in others, as the *Appendicularia* (Plate XVI., Fig. 3), the noto-cord and neural cord persist throughout the entire life of the animal. The life-history of these forms is of extreme interest; but, as it can be found in any text-book, we will pass to the feature that has rendered some of the class most conspicuous.

In exploring the depths of southern seas, among others we shall see a columnar form, the *Pyrosoma*, or "fire-body" (Plate XVII.), the giant of the Tunicates. It is an aggregation of individuals, forming a hollow cylinder closed at one end, and from two inches to four feet in length.[41]

The *Pyrosomæ* are richly tinted during the day; but at night, as their name implies, they resemble incandescent bodies. Humboldt refers to the spectacle he enjoyed when passing through à zone of them in the Gulf Stream, distinguishing by their light the forms of fishes, that, bathed by their gleams, stood out in bold relief far below the surface.

The light is extremely beautiful. That of the Atlantic forms is said to be polychroic, or an intense green; while in the very large species it is azure. So brilliant and striking is the light, that the impression is gained that it proceeds from the entire surface of the animal; but this is not the case, according to Panceri.[42] When the *Pyrosoma* is moving along in its curious fashion, — which calls to mind the old stern-wheel steamers, — and is undisturbed, the light is intermittent, now flashing from one cell, and now from another; the vast number of gleams giving it the appearance at times of constant light over the entire surface.

Panceri found that the luminous bodies produced an albu-

minoid substance that may become diffused by handling, and
retain its luminosity for some time. Curiously enough,
fresh water increases the intensity of the light, and causes it
to continue for a longer period. The intensity of the light
may be realized, when we learn from Figuier that Bibra, a
Brazilian navigator, employed six *Pyrosomœ* to illuminate his
cabin, which was thus rendered so bright that he could read
to a friend the description he had written of these living
lanterns.

Mr. Bennett, the naturalist, thus describes his experience
with these beautiful creatures: " On the 8th of June, being
then in latitude 30° south, and 27° 5' west longitude, having
fine weather and a fresh south-easterly trade-wind, and the
thermometer ranging from 78° to 84°, late at night the mate
of the watch called me to witness a very unusual appearance
in the water. This was a broad and expansive sheet of phosphorescence, extending from east to west as far as the eye
could reach. I immediately cast the towing-net over the
stern of the ship, which soon cleaved through the brilliant
mass, the disturbance causing strong flashes of light to be
emitted; and the shoal, judging from the time the vessel
took in passing through the mass, may have been a mile in
length. On taking in the towing-net, it was found half filled
with *Pyrosoma atlanticum*, which shone with a beautiful pale-
greenish light. After the mass had been passed through
by the ship, the light was still seen astern, until it became
invisible in the distance, and the ocean became hidden in
the darkness as before this took place.

" The second occasion of my meeting these creatures was
in a high latitude, and during the winter season; the
weather dark and gloomy, with light breezes from north-

north-east, in latitude 40° 30′ south, and 138° 3′ east longitude, at the western entrance to Bass's Straits, and about
eight o'clock P.M., when the ship's wake was perceived
to be luminous, while scintillations of the same light were
abundant all around. To ascertain the cause, I threw the
towing-net overboard, and in twenty minutes succeeded in
capturing several *Pyrosomœ*, which gave out their usual pale-
green light; and it was, no doubt, detached groups of these
animals which occasioned the light in question. The beautiful
light given out by these molluscans * soon ceased to be seen ;
but, by moving them about, it could be reproduced for some
length of time after. The luminosity of the water gradually
decreased during the night, and toward morning was no
longer seen."

M. Peron, says Figuier, observed the beauties of the
Pyrosoma atlanticum on his voyage to the Isle of France.
The wind was blowing with great violence, the night was
dark, and the vessel was making rapid way, when what
appeared to be a vast sheet of phosphorus presented itself,
floating on the waves, and occupying a great space ahead of
the ship. The vessel having passed through this fiery mass,
it was discovered that the light was occasioned by animalcules swimming about in the sea, at various depths, round
the ship. Those which were deepest in the water looked
like red-hot balls, while those on the surface resembled
cylinders of red-hot iron. Some of the latter were caught ;
they were found to vary in size from three to seven inches.
All the exterior of the creatures bristled with long, thick
tubercles, shining like so many diamonds; and these seemed

---

* When this account was written, the Tunicates were supposed to be
mollusks. — NOTE BY THE AUTHOR.

to be the principal seat of their luminosity. Inside, also, there appeared to be a multitude of oblong, narrow glands, exhibiting a high degree of phosphoric power. The color of these animals, when in repose, is an opal yellow, mixed with green; but, on the slightest movement, the animal exhibits a spontaneous contractile power, and assumes a luminous brilliancy, passing through various shades of deep red, orange green, and azure blue.

Professor Moseley captured a *Pyrosoma* four feet long, ten inches in diameter, with walls an inch in thickness. It was placed upon the deck of the vessel, and, when the naturalist wrote his name upon the animal with his finger, it came out in letters of fire: each letter seeming to increase in size, until the entire name was lost in a blaze of light, that radiated rapidly and soon suffused the entire animal; presenting a marvellous spectacle, and showing, in a striking manner, how intimately the animals are connected. In Plate XVII. a *Pyrosoma* of the largest size is shown in comparison with a native diver.

Sir Wyville Thompson observed the *Pyrosomæ* off the Cape Verde Islands, and refers to the "blaze of phosphorescence and train of intense brightness that followed the ship;" and, while he did not experiment with the animals in his cabin, as did Bibra, he says, "It was an easy matter to read the smallest print, sitting at the after port in my cabin; and the bows shed on either side rapidly widening spaces of radiance, so vivid as to throw the sails and rigging into distinct lights and shadows. The first night or two after leaving San Iago, the phosphorescence seemed chiefly due to large *Pyrosomæ*, of which we took many specimens in the tow-net, and which glowed in the water with a white light like that from molten iron."

Not the least wonderful feature of this animal is the variety of tints; white, green, various shades of deep red, orange green, and azure blue having been ascribed to it by different observers, — a fact that must stamp it as the most wonderful of all light-givers, a veritable living diamond.

One of the most remarkable exhibitions of phosphorescence was observed in January, 1880, by Commander R. E. Harris of the steamship "Shahjehan," — a display so unusual that I quote Capt. Harris's letter in full; and, while he is inclined to consider the exhibition as possibly electric, it would seem that the luminous objects referred to were phosphorescent animals of some kind, and possibly may have had some connection with the phenomenon.

" The most remarkable phenomenon," says Capt. Harris, " that I have ever seen at sea was seen by myself and officers on the 5th instant, between Oyster Reef and Pigeon Islands (Malabar coast). At ten P.M. we were steaming along very comfortably. There was a perfect calm, the water was without a ripple upon it, the sky was cloudless, and, there being no moon, the stars shone brightly. The atmosphere was beautifully clear, and the night was one of great quietude. At the above-named hour I went on deck, and at once observed a streak of white matter on the horizon bearing south-south-west. I then went on the bridge, and drew the third officer's attention to it. In a few minutes it had assumed the shape of a segment of a circle, measuring about forty-five degrees in length, and several degrees in altitude about its centre. At this time it shone with a peculiar but beautiful milky whiteness, and resembled (only in a huge mass, and greater luminous intensity) the *nebulæ* sometimes seen in the heavens. We were steaming to the southward; and, as the bank of light extended, one of its arms crossed our path. The whole thing appeared so foreign to any thing I had ever seen, and so wonderful, that I stopped the ship just on its outskirts, so that I might try to form a true and just conception of what it really was. By this time all the officers and engineers had assembled on deck to witness the scene, and were all equally astonished and interested. Some little time before the first body of light reached the ship, I was enabled, with my night glasses, to resolve in a measure what appeared to the unassisted eye a huge mass of nebulous matter. I distinctly saw spaces between what again appeared to be waves of light of great lustre. These came rolling on with ever-increasing rapidity till they reached the ship; and in a short time the ship was completely surrounded with one great body of undulating light, which soon extended to the horizon on all sides. On looking into the water, it was seen to be studded with patches of faint, luminous, inanimate matter, measuring about two feet in diameter. Although these emitted a certain amount of light, it was most insignificant when compared with the great waves of light that were floating on the surface of the water, and which were at this time converging upon the ship. The waves stood many degrees above the water, like a highly luminous mist, and obscured by their intensity the distant horizon; and, as wave succeeded wave in rapid succession, one of the most grand and brilliant, yet solemn, spectacles that one could ever think of was here witnessed. In speaking of waves of light, I do not wish to convey the idea that they were mere ripplings, which are sometimes caused by fish passing through a phosphorescent sea; but waves of great length and breadth, or, in other words, great bodies of light. If the sea could be converted into a huge mirror, and thousands of powerful electric lights were made to throw their rays across it, it would convey no adequate idea of this strange yet grand phenomenon.

" As the waves of light converged upon the ship from all sides, they appeared higher than her hull, and looked as if they were about to envelop her; and, as they impinged upon her, her sides seemed to collapse and expand.

"Whilst this was going on, the ship was perfectly at rest, and the water was like a millpond.

"After about half an hour had elapsed, the brilliancy of the light somewhat abated, and there was a great paucity of the faint, lustrous patches which I have before referred to; but still the body of light was great, and, if emanating from these patches, was out of all proportion to their number.

· "This light I do not think could have been produced without the agency of electro-magnetic currents exercising their exciting influence upon some organic animal or vegetable substance. And one thing I wish to point out is, that, whilst the ship was stopped and the light yet some distance away, nothing was discernible in the water; but, so soon as the light reached the ship, a number of luminous patches presented themselves: and, as these were equally as motionless as the ship at the time, it is only natural to assume that they existed, and were actually in our vicinity, before the light reached us, only they were not made visible till they became the transmitting media for the electro-magnetic currents. This hypothesis is borne out by the fact that each wave of light in its passage was distinctly seen to pass over them in succession; and, as the light gradually became less brilliant, they also became less distinct, and had actually disappeared so soon as the waves of light ceased to exist."

A little Ascidian called the *Salpa* is quite famous for its luminous properties. Like the previously mentioned form, it is a free swimmer, two kinds of individuals being recognized. One is known as solitary; while the others are termed chain zoöids, being many joined together, forming long chains, the links represented by the individual animals.

The *Salpa spinosa*, a familiar form upon our coasts, is quite
cylindrical, often a little flattened above and below, and
seemingly moulded in glass, so beautiful is its structure upon
examination. As small and common as they are, they have
created much discussion. Some observers deem their development one of the most remarkable instances of the alternation of generations. Chamisso, the German poet-naturalist,
explains the relationship as follows: "A *Salpa* mother is not
like its daughter or its own mother, but resembles its sister,
its granddaughter, and its grandmother." Dr. W. K. Brooks
has given much attention to these forms in this country;
and, from his point of view, the alternation of generations
would be impossible.

The *Salpæ* give little signs of animation. "The only conspicuous vital action," says Professor Owen, "is the rhyth-
mical contraction and expansion of the mantle, in which the
elasticity of the outer tunic antagonizes the contraction of
the inner one. During expansion, the sea-water enters by the
posterior aperture, and is expelled, in contraction, by the anterior one; its exit by the opposite end being prevented by a
valve. The re-action of the jet, which is commonly forced
out of a contracted tube, occasions a retrograde movement
of the animal." As they move along, on dark nights, they
present the appearance of fiery serpents or luminous ribbons
(Plate XVI., Fig. 1), winding their way over the sea,—a
most striking spectacle.

The light of *Salpæ* observed by Giglioli was confined to
the so-called nucleus, but was not constant; indeed, some
were luminous and some were not. This was particularly
evident in the month of September, when the exploring-ship
"Magenta" passed through a bed of these little creatures,

PLATE XVI.

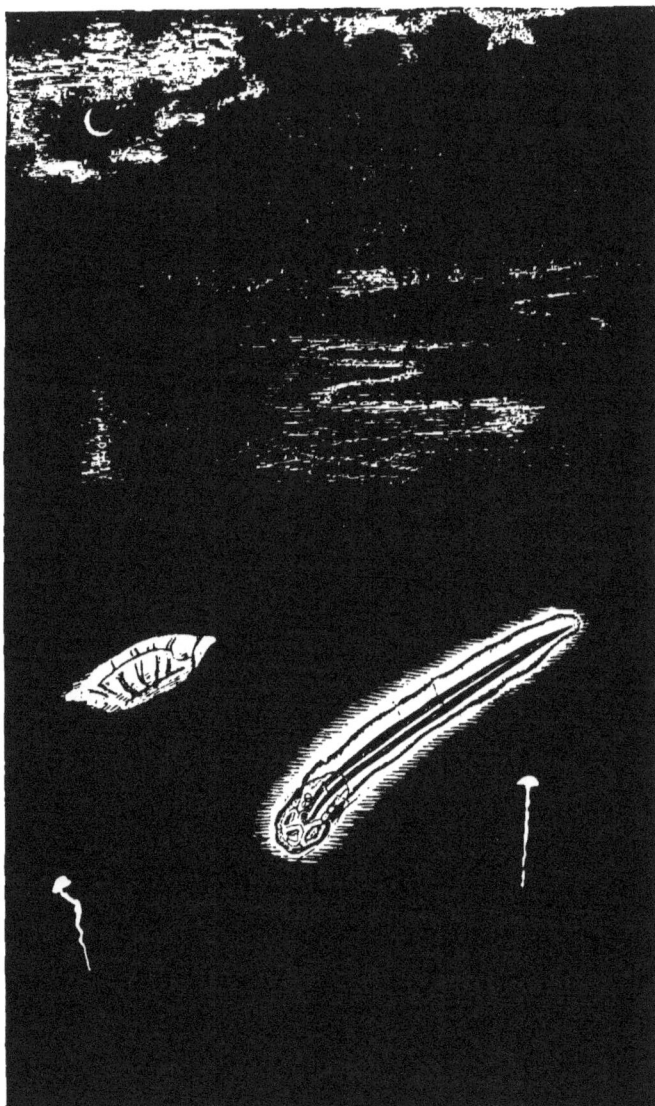

CHAIN OF SALPS

Salpa spinosa   *Appendicularia*

fifteen miles in extent. Some observed in the South Atlantic had the nucleus tinged with a brilliant red light. Very similar to *Salpa* is *Doliolum*, which seems to burn with a vivid green light scattered over the entire body, and is one of the emeralds of the sea. In the very lowest order (*Copelatæ*) of the Tunicates, we find an interesting, indeed remarkable, light-giver, the *Appendicularia* (Plate XVI., Fig. 3). It resembles a tadpole with quite a long tail, retaining in its adult life features that only characterize the *larvæ* of others of the group. Professor Agassiz has noticed two specimens on the New-England coast, and they are very common in both tropical and temperate waters of various regions.

Some of the species are veritable house-builders, forming a gelatinous protection covering called a test. This habitation, if so we may term it, is formed or secreted with considerable rapidity, and is quite an elaborate affair; having two front chambers and a middle one large enough for the tail to move with ease. Curious to relate, this transparent residence is, according to Filhol, only used a few hours, being then deserted and another formed; so that its life would seem to be spent in making houses and deserting them.

The light of certain *Appendiculariæ* is almost as remarkable as that of the *Pyrosomæ*, in the variety of its coloring; one, according to Giglioli, appearing first red, then blue, and finally green. The seat of the luminosity, which appears in intense flashes, was the central axis of the tail, or caudal appendage. Between Montevideo and Batavia in the South Atlantic, this naturalist observed many of these little creatures, nearly all of which showed these tri-colored favors; and in the Indian Ocean some were seen emitting white, blue, and green lights, marking them as among the most striking of all the light-givers.

Charles William Peach, an English naturalist, has observed the tadpole form of *Cynthia* to emit light. *Cynthia pyriformis* resembles a peach in form, size, and even bloom; its tests having rich reddish tints. It is a familiar form in deep water from Cape Cod to Greenland and across to Scandinavia. It is one of the most common objects on our New-England beaches after storms. The heavy seas throw it up from its hiding-places. To the student or interested visitor it is a beautiful object.

CHAPTER XII.

FINNY LIGHT-BEARERS.

IF it were possible for human beings to penetrate to the abyssal depths of the ocean, finny torch-bearers would be found from the very surface to nearly four miles beneath it; existing in many cases under conditions almost incomprehensible when the enormous attendant pressure is considered. While it is extremely difficult to tell the exact depth from which a fish is taken by the dredge, sufficient data has been secured for naturalists to assume, though there is great difference of opinion, that, to a greater or less degree, the forms of certain depths have certain peculiarities. These are often seen in the organs of vision, which have been modified in many ways by the lack of light. Thus the eyes of forms that are found living five or six hundred feet below the surface are often extremely large, as in *Beryx* (Plate XVIII., Fig. 1), as if to absorb the faintest beams of sunlight that may penetrate this distance. As we descend to twelve hundred feet, the eyes seem to grow larger; and beyond this, large and small eyed fishes are found indiscriminately. The former evidently use these organs; while those with small eyes are provided with remarkable organs of touch, — long feelers which can be thrown forward, or moved to a more or less extent, and used as the blind man uses his cane. An interesting, indeed remarkable, example of this is seen in the fish *Bathypterus longipes*, Günther found at a depth varying from one-half to three-quarters of a mile from the surface in the Atlantic. The eyes are extremely small, apparently useless; but the blind man's cane is here, as the pectoral fins are modified to serve as feelers, two rays almost as long as the entire fish extending from the back of the head. As the fish swims freely, the fins are trailed behind; but, does it approach a prospective victim, the articulation of these wonderful feelers enables them to be thrown forward as a cane in advance of the fish. They are divided at the tip, and form a delicate sense-organ with which to explore the depths of this abyssal world. Upon the ventral fins, there are two similar rays, that serve a like purpose.

Many fishes having remarkable feelers have quite recently been discovered, and among them *Eustomias obscurus* (Plate XIX.), a fish found at a depth of twenty-seven hundred meters, which has a long tentacle dependent from the lower jaw.

When we penetrate beyond a certain depth, we find blind fishes as well as those possessing eyes; and all the forms of the greater depths are adapted to their life under the consequent enormous pressure in a remarkable manner. The bones are friable and cavernous, and loosely connected. Many are covered with a thick mucus, while many more have curious plates, that are so many torches or lanterns to emit light for their possessors. As some of the fishes have eyes and no phosphorescent organs, while others are luminous and perhaps blind, and knowing that all are carnivorous, we may well imagine that a fierce struggle for existence is carried on in this distant world of the sea. The lamps of

some forms must attract their enemies; while, on the other
hand, they may constitute a lure, dazzling weaker forms,
which fall victims to their curiosity.

Among all the light-givers, these deep-sea lantern-bearers
are the most interesting, and typical of the mysterious realm
from which they are taken by the ingenious inventions of
mankind. Some are luminous over their entire surface, as
the *Harpodon*, or Bombay duck (Plate XVIII., Fig. 2).
Others have a series of plates extending along the side, that
resemble the open ports of a steamer. Some possess gleaming head-lights, the locomotives of the sea; while others have
their lights confined in groups.

While the expeditions of the "Challenger," "Talisman,"
"Albatross," and "Travailleur" have resulted in the discovery of what seems a remarkable presentation of these
light-givers, we can well imagine, understanding the difficulties of deep-sea dredging, that the largest and perhaps
most interesting of these forms are yet undiscovered, and
that the greatest mysteries may never be revealed. The
difficulties that attend, and the chances against, the capture
of deep-sea fishes, can be perhaps realized by my young
readers, if they imagine a large balloon sailing along over
the country at an elevation of from four to five miles, dragging a dredge ten or twelve feet wide. Few active boys
or girls would be caught by such a device; only the sluggards that were fast asleep would be trapped. The comparatively small dredge at the end of a six-mile rope, dragging
along and creating an unusual commotion in the silent submarine world, secures only a few forms, the sluggards and
mud-lovers, as a rule: so that fishes taken at extreme depths
are prizes indeed. The "Talisman" took the fish *Bythites*

crassus from a depth of about two miles. The naturalists of the " Challenger " expedition captured the *Bathyophis ferox* about three miles from the surface, or, to be exact, five thousand and nineteen meters. The American exploring-vessel, the " Albatross," under the direction of Professor Spencer F. Baird, has exceeded any of these hauls; in 1883 making a capture of five species in twenty-nine hundred and forty-nine fathoms.

While luminous fishes have been known for many years, the " Challenger " expedition brought many new forms to light, and the work accomplished by her officers may be said to have given a new impetus to the study of deep-sea forms. Off the north-west coast of Australia, the " Challenger's " trawl captured the curious black fish *Echiostoma microdon.* The luminous spots were few in number, but so arranged as to be of the greatest service: thus two are found just below the eyes; above the maxillary there is a narrow, elongated one, with a smaller spot nearer the eye. *E. micripnus*, found in twenty-one hundred and fifty fathoms, has long, fringed barbels, and small, round luminous spots above the maxillary, resembling a rudimentary eye.

Referring to this interesting torch-bearer, Dr. Günther says, " The fishes of the family *Stomiatidæ*, to which this genus belongs, are armed with formidable teeth, — a certain indication of their predaceous habits and voracity. Their long body is covered with a smooth, scaleless skin, of an intensely black color. The vertical fins are close together, near the end of the tail, as in the pike, forming a powerful propeller, by a single stroke of which the fishes are enabled to dart with great rapidity to a considerable distance. A long filament is suspended below the chin; and, as it is frequently fringed at its extremity, it evidently serves as a lure for other fishes or *animalculæ*. Series of luminous, globular bodies run along the lower half of the body and tail; and some others of larger size occupy the side of the head, generally below the eye or behind the maxillary bone. This fish is sixteen inches in length. The end of the barbel, which was thickened, was flesh-color with a rose tint; there was also a rose tint on the dorsal and anal fins. The rest of the animal was of a dark color. The phosphorescent spots along the belly and radial and lateral line were red, as was also that below the eye."

It is not often that the light of these fishes is seen; but the late Professor Willemoes Suhm, while watching the great trawl come over the side upon a calm night, noticed a gleaming spot, and taking it out found it to be the little fish *Sternoptyx*. In referring to it he says, "It hung in the net like a golden star, as it came out of the darkness."

As the *Sternoptyx* is a delicate little creature, and quite defenceless, its illumination must be a fatal gift. This is equally true of the *Argyropelecus hemigymnus* (Plate X., Fig. 8), a curiously formed fish, — deep in the body, tapering off suddenly to the tail, as if a piece had been bitten out by some large fish. Referring to the figure, it will be seen that the luminous organs are grouped; four being at the side of the tail, six midway between it and the line of the dorsal fin, and many others around the edge of the ventral surface, — one hundred and six in all: so that if all these plates are luminous, the *Argyropelecus* must present a dazzling sight as it darts along in the blue waters of the Mediterranean, where it has been most commonly observed.

Concerning the functions of these organs, there is still

much controversy. The opinions of Ussow, Leydig, and others will be found in their papers referred to in the bibliography; and, as the question is thoroughly a technical one, its further discussion is omitted. As early as 1865 Professor Leuckart suggested that the curious plates (Plate X., Fig. 4) were organs of sight, or accessory eyes. In 1879 Dr. M. Ussow, of the University of St. Petersburg, gave the world an account of his researches upon the plates of the genera *Astronesthes, Stomias* (Plate XX.), *Chauliodus* (Plate XXI., Fig. 4), *Scopelus* (Plate X., Fig. 1), *Maurolicus, Gonostoma,* and *Argyopelecus,* small fishes, most of which were found in the Mediterranean. This was followed by similar investigations by Dr. Leydig of Bonn, and Dr. Günther.

A well-known phosphorescent fish is seen in *Scopelus,* which bears upon its sides and various parts of the body numbers of spots (Plate X., Fig. 1), which, if all luminous, mark it as one of the most brilliant of the light-givers. The appearance of these organs in reflected light is shown in Plate X., Fig. 2.

The snake-like *Stomiasboa* (Plate XX.), from a depth of twenty-seven hundred feet, is perhaps the most hideous of the light-givers; its large mouth and ferocious teeth giving it a bull-dog aspect, which in a large fish would make a veritable dragon. But *Stomias.*is not over twelve or fifteen inches in length, though quite large enough to terrify the smaller fry. The specimen figured was taken in the Gulf of Gascony by the naturalists of the " Talisman," from its home, a mile and a quarter beneath the surface. The sides of the body are provided with a double row of luminous disks, which, according to M. Filhol, " cause the fish to be surrounded by a brilliant luminous aureola."

PLATE XVIII.

CHIASMODUS.   SUN-FISH.   PLAGIODUS.   HARPODON.

BERYX.

In Plate XVIII., Fig. 4, is shown a large light-giver, — the *Plagiodus*, a fish six feet in length. According to Dr. Günther, it emits light from various parts of its surface; the tips of the fins gleaming with a soft phosphorescence similar to that of the large-eyed *Beryx* (Fig. 1) of same plate. The latter attains a length of about twenty inches.

Quite as ferocious in appearance as the *Stomias* is *Chauliodus* (Plate XXI., Fig. 4), with long, lance-like teeth, gleaming fins, and a row of small phosphorescent plates that perhaps sparkle like so many gems as their bearer sails along in the greater depths.

Exaggerations are often termed "fish-stories," for the reason perhaps that improbable tales are related concerning the denizens of the sea by fun-loving mariners; but the most remarkable stories that the vivid imagination of those who go down to the sea in ships has ever devised are not as remarkable as the simple truths regarding the every-day history of fish-life. What can be more astonishing than the fact that these delicate forms are enabled to live in water where the pressure is so great that hard wood is crushed and glass reduced to powder? If a decade or so ago a statement had appeared in the daily press, to the effect that a fish had been discovered which could swallow another five times its own bulk, it would in all probability have been classed as a "fish-story," — too big an one, indeed, to have even the merit of comical exaggeration: yet such a fish does exist in the black swallower, or *Chiasmodus* (Plate XVIII., Fig. 5); the fish, besides being luminous, possessing this extraordinary faculty. The jaws, by a special arrangement, are capable of great extension; so that the fish actually draws itself over its prey, that may be many times its own bulk. The skin of the

swallower seems to possess a rubber-like character, stretching
to enormous dimensions, and often, when filled with gas,
carrying the glutinous light-bearer into the upper regions
of the ocean.

Malacosteus niger, Ayres (Plate XXII.), is a rare fish,
from a depth of two-thirds of a mile; though several specimens have recently been taken by the United-States Fish-
Commission, and others by the "Talisman" off Morocco, in
forty-eight hundred feet of water. It is of small size, from
thirteen to fourteen centimeters in length, of a velvet-black
hue, and possesses two large luminous organs upon the head;
one of which, according to M. Filhol, who observed the light
in the living fish, emits a golden, and the other a greenish
phosphorescence. We have here, then, a fish that vies with
the *Appendicularia,* and other forms which we have seen
emitting light of more than one color. It is possible that
the rays of light from these spots project ahead of the fish,
in the manner shown in the accompanying figure, in which
the appearance of the light is of course conjectural; but as
to the meaning of the different colors, are they a system of
signals cunningly devised by Nature to enable *Malacosteus*
to distinguish its kind in the profound depths of the ocean,
or are they merely lures of more than ordinary brilliancy?

In some fishes the luminous organs are extremely small,
almost invisible to the naked eye, and often spread over a
large extent of surface. Such an instance is seen in *Eustomias obscurus* (Plate XIX.) and *Neostoma.* In the former,
an attenuated carnivorous fish of a jet-black color, we see
another example of remarkable feelers, or sense-organs.

While these forms are probably free swimmers, there are
many others that are mud-dwellers, of most extraordinary

make-up, literally living bags, or rather mouths. The *Mela-nocetus Johnstoni* (Günther, Plate XXIII.) is one of these; having an enormous pouch, with a fishing-rod upon its head similar to that of our common *Lophius*. *Melanocetus* probably buries itself in the ooze, as shown in the engraving, allowing the tip of its tentacle, or rod, to protrude; and, when the living bait is touched, it opens its cavernous mouth and seizes the victim.

Still more remarkable is the *Eurypharynx pelecanoides,* which has a mouth of enormous dimensions (Plate XXIV.), from which depends a pelican-like pouch. This form is interesting, from many peculiarities; among which may be mentioned the fact, that the bronchial arches are here simple bars, five in number, having no connection with the cranium. The mouth can open to a surprising extent, the lower jaw being composed of two pieces attached to the cranium by a movable joint, so that it swings literally in various direc-tions. The fish probably feeds by swimming along the bottom blindly, ingulfing various animals, holding them by its interlocked teeth. This phenomenal fish was taken in 1882 by Vaillant, the French scientist, twenty-five hundred metres from the surface; while another genus of these deep-sea, eel-like creatures was described in 1883 by Gill and Ryder, who called it *Gastrostomus bairdii.* In the latter, the jaw is six or seven times as long as the cranium.

One of the most striking phosphorescent fishes is a small shark, *Squalus fulgens,* also described and figured by Kner as *Leius ferox,* which, in general appearance, somewhat resem-bles the black or brown nurse (or *Scymnus*) of our Southern coast. This interesting light-giver was discovered by Dr. Bennett, and the following is his version of the find: "Being dark when I first saw it shining in the net, it resembled a *Pyrosoma*, emitting, as it did, a bright phosphorescent light. This was in latitude 2° 15′ south, longitude 163° west. The length of my specimen was five inches and a half. It is not a little singular that my brother, the late D. F. Bennett, obtained a specimen of this fish in the same latitude, and another in latitude 55° north, longitude 110° west. The first was taken in the daytime, and was ten inches in length, — much larger in size than my specimen. The second was taken at night, and its entire length was a foot and a half: both were alive when captured,· and fought fiercely with their jaws, tearing the net in several places. On placing my fish in sea-water, and observing . it in the dark cabin, it swam about for some time, emitting a bright phosphoric light; and when this had become so faint as to be almost imperceptible, it was readily rekindled on the animal being disturbed or excited. My specimen was of a perfectly black color, and died about four hours after it had been taken. The luminosity was retained for some hours after life was extinct.

"The form of the shark, as indeed its whole structure, is peculiar. It no doubt belongs to the subgenus *Scymnus.* My specimen having been accidentally lost, I am unable to give a minute description of it. My brother was more fortunate. I will, therefore, give his account of so novel and interesting a fish. The body is cylindrical, rather slender, and tapers finely towards the tail. Its prevailing color is dusky brown; a broad black band, or collar, passes around the throat; and the fins are partially margined with white (my specimen, being small and young, varied in this respect, being black, with the fins of a less intensity of color); the skin rough, as is usual in the shark tribe. The number of gill-apertures is five on each side. The fins are short, and for the most part disposed in a round form; the dorsals are two in number, small, and placed far back; the tail-fin is unequally divided, the upper being the longest and largest lobe. The head is flat; the snout prominent, rather pointed, and has two nostrils at its extremity. There is, also, on each side of the upper and back part of the head, a large oval orifice, like a spiracle or nostril, provided with a valve, and communicating with a corresponding aperture in the roof of the mouth. The mouth is capacious; and the dark skin around it is incised on each side to some extent beyond the commissure of the lips, exposing a white elastic membrane beneath. The upper jaw is armed with many rows of small, sharp teeth; while the lower has only a single row of perpendicular teeth, or, rather, an elevated plate of bone, sharply toothed on its summit, and bearing a close resemblance to a segment of the surgical circular-saw called a trephine. The eyes are much more prominent and dilated than is usual in sharks; the iris is black, the pupil transparent and of a greenish color.

" When the larger specimen, taken at night, was removed into a dark apartment, it afforded a very extraordinary spectacle. The entire inferior surface of the body and head emitted a vivid and greenish phosphorescent gleam, imparting to the creature, by its own light, a truly ghastly and terrific appearance. The luminous effect was constant, and not perceptibly increased by agitation or friction. I thought at one time it shone brighter when the fish struggled, but I was not satisfied that such was the fact. When the shark expired (which was not until it had been out of the water more than three hours), the luminous appearance faded

entirely from the abdomen, and more gradually from other
parts; lingering the longest around the jaws and on the fins.

"The only part of the under surface of the animal which
was free from luminosity was the black collar around the
throat; and while the inferior surface of the pectoral, anal,
and caudal fins shone with splendor, their superior surface
(including the upper lobe of the tail-fin) was in darkness; as
also were the dorsal fins, back, and summit of the head. I
am inclined to believe that the luminous power of this shark
resides in a peculiar secretion from the skin. It was my
first impression that the fish had accidentally contracted
some phosphorescent matter from the sea, or from the net in
which it was captured; but the most rigid investigation did
not confirm this suspicion, while the uniformity with which
the luminous gleam occupied certain portions of the body
and fins, its permanence during life, and decline and cessa-
tion upon the approach and occurrence of death, did not leave
a doubt in my mind that it was a vital principle, essential to
the economy of the animal. The small size of the fins would
appear to denote that this fish is not active in swimming;
and, since it is highly predaceous, and evidently of nocturnal
habits, we may, perhaps, indulge in the hypothesis that the
phosphorescent power it possesses is of use to attract its
prey, upon the same principle as the Polynesian Islanders
and others employ torches in night-fishing."

LUMINOUS FISH.
(Eustomias obscurus.)
From depth of 8,100 feet.

CHAPTER XIII.

FINNY LIGHT-BEARERS — (SURFACE FORMS).

ON calm nights the splash of the oars and the fall of spray from the bow of the boat startle many fishes resting at or near the surface, which dart away like comets, leaving a blaze of light behind, and giving the impression that they are light-givers or phosphorescent. This does not always follow; as, while many possessors of luminous spots undoubtedly approach the surface at night, as *Scopelus* (Plate X., Fig. 1), many owe their brilliant appearance to the luminosity of the medium in which they swim; in other words, the vigorous motion of their fins produces the same effect and result that is attained by darting the hand through water bearing phosphorescent animals. If such a display is produced by one fish, we may well imagine that a school moving rapidly would create a light of considerable intensity.

Drifting over a school of menhaden, and peering down among them, each fish seems outlined in a golden halo; while coruscations of light appear to flash from the fishes as they move along, the presence of the school being indicated upon the water by a pale luminous spot.

In more active fishes, as the mackerel, the display is still more brilliant, often presenting a blaze of light upon the surface, visible from the mast-head of a vessel for a long distance, and often resulting in the capture of an entire school; as the mackerel-men, aware of the light produced by the fish, keep a lookout in the foretop; and upon its discovery, the great net is passed around it, the fishes becoming victims to the light they inadvertently produce. When the mackerel are tossed into the boat, they roll over in a golden mass in their struggles, hurling a cloud of spray into the air over boat, net, and men. In handling these fishes, phosphorescent matter will sometimes come off upon the hands, and the gleaming fluid is seen running from the bodies; so, possibly, in some instances, the fishes possess a luminous secretion, as in the case of the shark of Dr. Bennett.

The sunfish (Plate XVIII., Fig. 3), an extremely common form on our eastern shores, appears to have a wide geographical range. In American waters, it is known as the sunfish, presumably from its oval shape. Two fins only are present, these being opposite one another, the tail represented by a mere ridge. The sunfish attains a height, from the tip of one fin to that of the other, of seven feet, and sometimes more, weighing several hundred pounds.[44]

Some years ago, while at the little fishing-village of Mayport, at the mouth of the St. John's River, Florida, one of these huge fishes ran aground on the bar, actually drawing too much water to cross. Its struggles attracted so much attention, that a boat was sent out, and the monster captured. I sent a photograph of the fish north, and the latter was afterwards purchased by the New-York Aquarium. It was the largest specimen of this fish I ever saw on exhibition.

So sluggish are they, that, at Ogunquit, Me., the fishermen frequently ran alongside of them as they rolled about at the surface, and, thrusting a boat-hook into the small mouth,

hauled them aboard; or, if too heavy, lashed them to the
side, in which position they were towed ashore, where the
liver, the only valuable portion, was secured; though the muscular tissue was sometimes appropriated by the boys of the
neighborhood, who found it a good substitute for India-
rubber as an interior for base-balls.

In a large specimen which I examined, the skin was covered with a remarkable mucilaginous envelope, in which were
numerous parasites; while in the mouth was a large goose-
barnacle, which was situated just far enough in to escape
being crushed by the formidable teeth. If asked to select a
fish showing evidences of possible phosphorescence, I should
name the sunfish, as the curious envelope of mucus seems
particularly adapted as the seat of this remarkable phenomenon; but I have not only never observed its luminosity,
but have been unable to obtain a direct statement from any
one in this country as to its light-emitting quality. I give
it a place among the luminous fishes, on the authority of
T. Spencer Cobbold, M.D., F.L.S., who says, in referring to
it, "It is nearly circular in form; and the silvery whiteness
of the sides, together with their brilliant phosphorescence
during the night, has obtained for it, very generally, the
appellation of sun or moon fish."

Karl Semper, in his "Animal Life," says: "The fishermen
of Nice assert that the moonfish (*Orthagoriscus mola*) is
luminous;" but as no scientist, that I am aware of, makes
a definite statement of personally observing its light, we
will leave the moon or sun fish among the forms which are
possibly phosphorescent, yet not proven so.

Statements are often made regarding the phosphorescence

of whales and other cetaceans; but the wondrous displays which they undoubtedly produce as they rise, perhaps to escape the ferocious attacks of the killer, are due only to the myriads of small light-givers, — *medusæ*, *salpæ*, crustaceans, and others, — which when disturbed become luminous.

Among the well-known phosphorescent fishes, the *Scopelus*, found in the greater depths, rises at times, at night, to the surface. *Scopelus humboldtii* (Plate X., Fig. 1), has a double row of luminous spots on each side of the abdomen. One of the spots, enlarged in reflected light, is shown in the same plate.

The phosphorescence of *Myctophum crenulare*, an ally of *Scopelus*, has been observed; and, at least on the Pacific coast, this little fish probably rises to the surface, a specimen an inch and a half in length having been taken from the stomach of an albicore (*Orcynus alalonga*) in the Santa Barbara Channel. In this specimen a phosphorescent spot was seen on each mandible near the symphysis, thirty-three along the abdomen, six in front of the ventral fins, six more between the latter and the origin of the anal, and twenty-one between the front of the anal fin and the base of the tail; quite enough, if all are luminous, to outline the little creature in lines of vivid brightness.

The long, arrow-like gars are peculiarly surface forms, it being evidently only with extreme difficulty that they leave the surface. Allied to them is *Hemiramphus*, in which the lower jaw only is elongated; and, according to Günther, this interesting fish has a gleaming phosphorescent pustule at the tip of its tail, a circumstance that makes it one, not only of the most unique of the surface forms, but of all the finny light-bearers. Many other forms known to possess luminous spots undoubtedly visit the surface at night, just as many large predatory fishes then come well in shore. Indeed, the night is the feeding-time of the southern fishes; at least, the season when they are upon their travels.

At Tortugas, on the Florida reef, the shoal to the west of the key was deserted during the day, except by schools of mullet, small barracuda, and a few others; but at night the sandy shoal seemed fairly alive with large fishes. Man-eaters, ten or fourteen feet long, ranged up and down, readily taking the hook; and nearly all the large fishes, which by day lived upon the outer reef or in the channel, could be taken here; while loud splashes and vivid displays of phosphorescence told that the large rays, indeed the great manta itself, ventured in shore in nocturnal rambles.

CHAPTER XIV.

LUMINOUS BIRDS AND OTHER ANIMALS.

IN floating over the great coral reef of the Florida penin-
sula one day, the boat startled a number of large cranes
which were standing upon a small key; and, as they laboriously
flew away, my companion, a sportsman of experience, related
to me the following incident: "Some years ago," he said, "I
was much more confined than I am at present, and rarely
had an opportunity of enjoying hunting during the day-time;
so I began a series of moonlight excursions about the reef,
generally securing a green turtle, if nothing else, and occasionally a large bird.

"One evening I visited one of the large keys; and before
I was ready to return the moon had gone down, leaving me
in the dark. It was a perfectly calm night, not a ripple
appearing upon the water, so that every sound was heard
with striking distinctness; and the break of the sea upon the
outer reef came to me in a sullen roar, occasionally varied by
the crash of some huge fish as it left the water. I was making
my way to my boat, when suddenly I perceived on the sands
several dim lights. Thinking it the reflection of the stars
upon the water, perhaps, I pushed on; and when I was
almost upon them, there came a flapping of wings, while
above I saw indistinctly the forms of several large cranes,

LUMINOSITY OF HERON'S BREAST.

that made their escape before I thought of shooting. The light disappeared with them; and my opinion is, that what I saw was phosphorescent light upon the breast of the birds."

I have been told by several sportsmen that they have heard of such an occurrence; and I have always been impressed with the belief that the greasy, oily, powder-down patches might become luminous under certain conditions, but never until the present year have been able to find reliable personal testimony. The following statement, prepared for me by Mr. Isaac W. Worrall of Philadelphia, shows that the phosphorescence of birds is a fact. To obtain a full account of Mr. Worrall's observations, I made out a list of questions, which he has kindly answered; and which, from the great interest connected with the occurrence, are given in full : —

"Upon what birds did you observe the luminosity?"

"The night heron (*Nyctiardea .grisea*) and blue crane (*Ardea cærulea*)."

"What was the situation of the light or lights?"

"One on the breast, and one on each side of the hips, between the hips and the tail."

"Upon how many birds did you observe the light?"

"Upon four different birds, including the one I shot."

"How far could you see the lights in the living bird?"

"I saw the light plainly at a distance of about fifty yards."

"Did you notice the reflection of the light upon the water?"

"No."

"Was the light brilliant enough to make a reflection?"

"Before I fired, the light appeared equivalent to two candles."

"Where was the bird you shot when first observed?"

"Standing in about six inches of water."

"Give a practical example of its brilliancy."

"When I aimed, I considered it equal to the light of a hand-lamp or lantern, and could see my gun-sight quite plainly against it."

"Could you have read by the light as it appeared when you took the bird from the water?"

"I have read small print with a dimmer light than that upon the bird immediately after it was shot."

"Do you think the bird can conceal or display its light at will?"

"I know the bird has full control of the light. I saw it open and shut it four times when I was crawling towards it. I stopped when it put out the light, and advanced when it was displayed again.". (The bird may have turned. — AUTHOR.)

"What was the state of the weather when you shot the bird?"

"A clear, dark night in spring." (Kansas.)

"Did you notice the sex of the bird?"

"No."

"How long did the light last after you shot the bird?"

"The light faded as the bird died, disappearing at death."

"Did you notice any odor while the light was apparent?"

"No."

"Did the luminous matter come off upon your hands?"

"I did not touch it."

" Was the light a steady glow ? "

" It lasted about as long as I could count twenty at moderate speed."

" What was the color of the light ? "

"It reminded me of phosphorescent wood, and was whitish."

When my informant first observed the light, he was a hundred and fifty feet away, and while slowly creeping toward it saw it disappear four times, the intervals between the disappearance and re-appearance being long enough for him to count twenty at a moderate rate ; from which he assumed that the bird has the light more or less under control, and governs it by raising or depressing the feathers that cover the powder-down patches. When he fired at the bird, the light on the breast was so intense that he distinctly saw the sight of his gun against it, and he describes its brilliancy as comparable to that of a lantern or hand-lamp. He did not notice a reflection upon the water, as he was some distance away, and in a recumbent position, which rendered it impossible. The bird fell where it was standing, in six inches of water; and taking it by the wings, he threw it upon the shore, noticing and watching the three phosphorescent spots, one in front, and one on each side of the hips, between the hips and the tail. The bird died slowly, *the light gradually dying out, and disappearing entirely with death ;* a fact which I consider to be of the greatest interest, showing that the phosphorescence is not an accidental occurrence, depending upon a favorable condition of the greasy powder-down patches, or associated entirely with their decomposition, but is essentially due to some physiological action, and dependent upon the life of the bird; and the areas of the powder-down patches may be considered true photogenic structures. The bird shot and examined by Mr. Worrall was known to him as the blue crane, and I assume from his description that it is the *Ardea cœrulea*. The other birds in which the light was observed were night herons. The light was in the so-called powder-down patches, which form a characteristic feature of the herons, and doubtless serve the same purpose, as a lure, in all.

In a night heron, which I recently obtained from a valley among the foothills of the Sierra Madre range, there were three of those patches, and any heron will show them. One is directly in front upon the breast, while the other two are upon each side, midway between the base of the tail and the upper portion of the thigh-bone. They are not visible unless the feathers which cover these portions are brushed aside, when a mass of oily small plumes are seen, of a decided yellow hue, growing closely together, and about two inches in length. A yellow powder will be found profusely mixed among them, and is due to their barbed tips breaking off as fast as they develop.

In my specimen, just after death the patches were quite oily, the substance coming off upon the hands, and smelling like ordinary bird oil. As soon as possible I took the bird into a perfectly dark room, to test it for phosphorescent light, but not the slightest gleam was perceptible. Just under the patches a large accumulation of fat is seen; and from these portions probably exudes the substance, which, during the life of the bird, becomes luminous upon exposure to the air. In the specimen alluded to, after it had been dead for several days, the shafts of the feathers of the patch seemed suffused with a dark oily substance. The feathers of the powder-down patches did not burn more readily than feathers from other parts, and the odor was the same.

These patches are not strictly confined to cranes and herons. The kirumbo (or *Leptosmus discolor*) of Madagascar has a highly developed patch upon each side of the rump. These birds are related to the rollers, and are remarkable for their games in mid-air. The bitterns have two pairs of powder-down patches, the true herons three, and the curious boatbills (*Cochlearius*) four pairs, which, if all luminous, must render them the centre of attraction in the South-American swamps.

The interesting oil-bird *Podargus* (or Guarcharo), that builds in the island of Trinidad and on various parts of the South-American coast, is a fruit-eating, nocturnal bird allied to the night-hawks. Curiously enough, it has no oil-glands, but two large powder-down patches, one on each side of the rump, composed, according to Dr. Sclater, who made the discovery, of about forty feathers each. In Plate XXVII., an ideal view is given of the possible appearance of the light of a large heron (*Ardeomega goliath*) of Africa.

Whether these lights are of sufficient brightness to attract fishes is a question; but, knowing that fishes are readily attracted by light of fire, we may well imagine that a crane or heron, if standing in the water in perfect stillness, with this soft light a short distance above it, might possibly avail itself of such a lure, though such a view is purely conjectural. Mr. Charles Harris of Pasadena, Cal., informed me that several years ago he entered a heronry in Maine on a dark night, and distinctly observed numbers of lights too large for insects; and, moreover, they disappeared with the birds, so that he was impressed that there was some association between the light and the herons.

That birds should be luminous is not, perhaps, strange. Other vertebrates appear to possess this gift in an equally remarkable manner. Some years ago an English gentleman, a lover of sport, was travelling in South America; and among the tales that he heard from the natives was one that related to a monkey with fiery eyes, as they expressed it. It seemed that one season, when the tribe was far up the branch of a small river, a woman wandered off into the forest at night, and returned much alarmed, stating to the rest that an animal had appeared to her with eyes gleaming like coals. Several of the natives went to the spot designated, and were repaid with a glimpse of the strange creature.

Such a tale was, of course, not received in good faith, being considered an example of the inventive fancy of these children of the forest; yet, curiously enough, Reninger the naturalist, who travelled extensively in Paraguay, states that he has seen the eyes of the monkey, *Nyctipithecus trivirgatus*, so brilliant in complete darkness that they illuminated objects at a distance of half a foot. In several instances I have referred to the phosphorescence of animals being used possibly as a warning; at least, this is the explanation given the phenomena by some observers, and one of the most interesting cases that may possibly come under this head is the luminosity of frogs' eggs. This has been noticed in various parts of Europe; masses of luminous matter being found about ponds and damp places, and termed *mucilage atmosphérique*, as it was believed by the simple peasants to be part of the tail of comets.

On one occasion several peasants were travelling from one

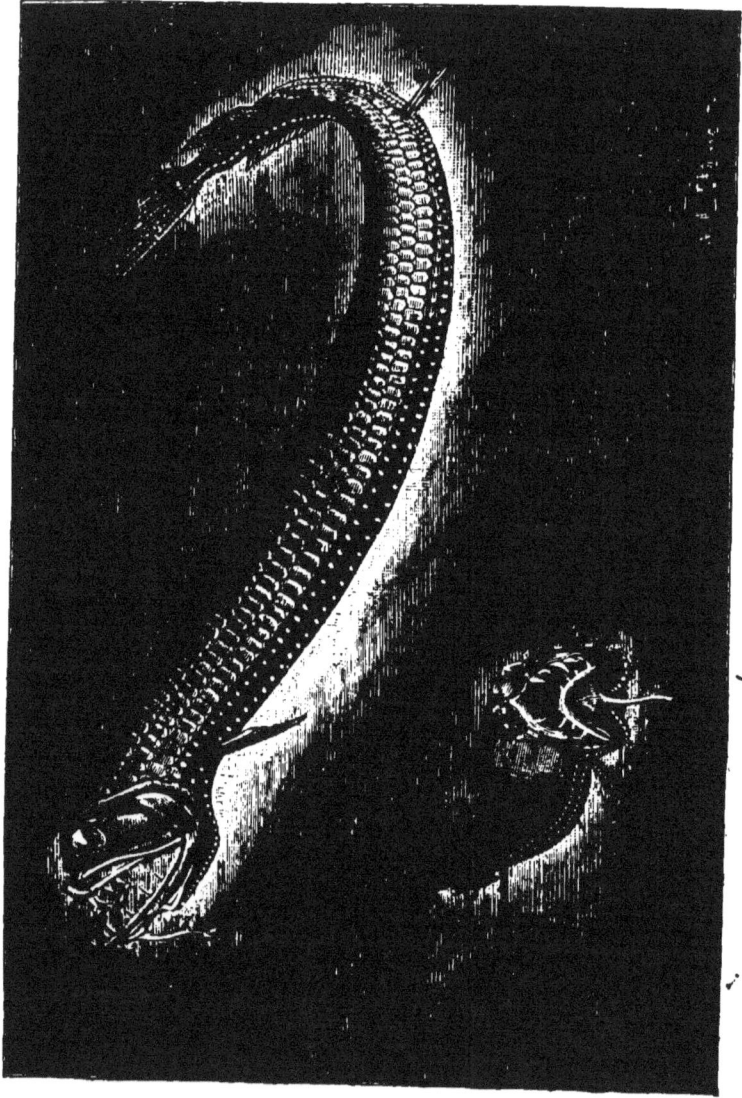

LUMINOUS FISH.
(*Stomias boa.*)

village to another at night, when suddenly a large meteor shot across the heavens, seeming to fall before them. A few miles farther on, in crossing a small swamp, they found several patches of a jelly-like matter, which gleamed as if at a white heat, which so alarmed them, that they ran into the next village, crying that a comet had fallen, and was burning up the earth. So much excitement was created that some scientific men visited the spot, finding the comet to be merely the mucus that had surrounded the eggs of a frog, and had become luminous. If the mucus was luminous when it surrounded the eggs, we may well imagine that birds would be deterred from eating them; but the luminosity probably precedes decomposition in the mass after the young have escaped.

Among the lizards, a gecko has been mentioned as a light-giver, as if these curious creatures were not remarkable enough in themselves without this attendant phenomenon. According to Dr. Carpenter, the eggs of the gray lizard have been seen to emit light; and in Surinam he states that a frog or toad is luminous, especially in the interior of its mouth. Thus we see that this strange light is found in some form from the lowest to the highest animals, — one of the commonest of phenomena, yet presenting a problem defying solution.

CHAPTER XV.

MAN'S RELATIONS TO THE PHENOMENON OF PHOSPHORESCENCE.

D R. PHIPSON, the eminent scientist, states that he once observed certain phenomena in man, the light being a brilliant scintillation of a metallic pink color.

It is well known that human beings under certain physical conditions become luminous. In some cases among the ignorant great excitement has been occasioned, and the victim avoided as a pest, or something capable of dire disaster to the entire community.

In a small German village, an English physician discovered a man who was luminous at night, and who had caused much alarm among the superstitious.

Bartholin records an instance of an Italian lady whom he calls *Mulier splendens*, who suddenly found that, when rubbed with a linen cloth in the dark, her body gave out a brilliant phosphorescent light; so that she appeared in a darkened room like a veritable fire-body, an awe-striking object to her superstitious servant, who fled from her speechless with fear and amazement, thinking that her mistress was being consumed.

Dr. Kane records a very curious instance of luminosity, probably electric, which played about his person. He was on his way with Petersen to an Esquimau settlement, in order to procure food. Their thermometer indicated 42° C. (44° Fahr). With their weary dogs and sledges, they had reached some untenanted huts at a place called Anoatok, after thirty miles march from the ship. " We took to the best hut," says Dr. Kane, " filled in its broken front with snow, housed our dogs, and crawled in among them. It was too cold to sleep. Next morning we broke down our door, and tried the dogs again. They could hardly stand. A gale now set in from the south-west, obscuring the moon, and blowing very hard. We were forced back into the hut; but after corking up all the openings with snow, and making a fire with our Esquimau lamp, we got up the temperature to 30° below zero, Fahr., cooked coffee, and fed the dogs freely. This done, Petersen and myself, our clothing frozen stiff, fell asleep through pure exhaustion; the wind outside blowing death to all that might be exposed to its influence. I do not know how long we slept, but my admirable clothing kept me up. I was cold, but far from dangerously so, and was in a fair way of sleeping out a refreshing night, when Petersen woke me with, ' Captain Kane, the lamp's out.' I heard him with a thrill of horror. . . . Our only hope was in relighting our lamp. Petersen, acting by my directions, made several attempts to obtain fire from a pocket-pistol; but his only tinder was moss, and our heavily stone-roofed hut or cave would not bear the concussion of a rammed wad. By good luck I found a bit of tolerably dry paper, and becoming apprehensive that Petersen would waste our few percussion-caps with his ineffectual snapping, I determined to take the pistol myself. It was so intensely dark that I had to grope for it, and in so doing touched his hand. *At that instant the pistol became distinctly visible. A pale-bluish light slightly tremulous, but not broken, covered the metallic parts of it,* — the barrel, lock and trigger. The stock, too, was clearly discernible, as if by the reflected light; and *to the amazement of both of us, the thumb and two fingers with which Petersen was holding it,* the creases, wrinkles, and circuit of the nails, clearly defined upon the skin. *The phosphorescence was not unlike the ineffectual fire of the glow-worm. As I took the pistol, my hand became illuminated also, and so did the powder-rubbed paper when I raised it against the muzzle.* The paper did not ignite at the first trial; *but the light from it continuing,* I was able to charge the pistol without difficulty, rolled up my paper into a cone, filled it with moss sprinkled over with powder, and held it in my hand whilst I fired. This time I succeeded in producing flame, and we saw no more of the phosphorescence. . . . Our fur clothing and the state of the atmosphere may refer it plausibly enough to our electrical condition."

Mr. James Moir of Saroch, Scotland, relates an equally strange personal experience, possibly connected with the electrical condition of the atmosphere. "In February, 1882," he says, "this part of Scotland was visited by a furious gale of wind, rain, sleet, and hail. The gale subsided considerably about five o'clock in the afternoon. At eight o'clock the sky was fairly clear, when a black cloud sprang up in the north, and the night became suddenly intensely dark. With the darkness came a tremendous shower of hail. All at once I was startled by a vivid flash of lightning close at hand, but without thunder. At the same instant I found myself enveloped in a sheet of pale, flickering, white light. It seemed to proceed from every part of my clothes, espe-cially on the side least exposed to the hail; and more particularly and brightly from my arm, shoulder, and head. Though I turned about pretty smartly, and shifted my position, I found it impossible to shake off the flickering flames. When I walked on they continued with me for two or three minutes, disappearing only when the violence of the blast was somewhat diminished. I felt no unusual sensation beyond the stinging of the hail, and no sound except that of the storm."

The adventures of John Stewart, who for many years drove a mail-gig between Dunkeld and Aberfeldy, Scotland, as given by an English paper, are well worth recording. On an extremely dark night, he and another man, climbing a rocky, heathery height in Rannock, were all at once set on flames by some mysterious fire, which appeared to proceed from the heather which they were traversing; and the more they tried to rub the flames off, the more tenaciously they seemed to adhere, and the more the fire increased in brightness and magnitude. Moreover, the long heather, agitated by their feet, emitted streams of burning vapor; and for the space of a few minutes they were in the greatest consternation. They believed that they barely escaped a living cremation. Of course their liberal share of native superstition, and the gloom of the night in the weird wilderness remote from human habitation, rendered their position the more alarming.

A wonderful phenomenon is noted by a gentleman living in Cheltenham, England. He was returning from Great Yarmouth to his house, a distance of three miles, and took the road of the Denes, intending to cross by the lower ferry. Before reaching it, a dark cloud coming from the south-east, off the sea, suddenly surprised him, and drenched him with

rain.   He jumped into the boat, and when the boatman had
pushed off, he remarked that every drop of rain hanging
from his hair, beard, and clothes was luminous with white
light, well seen, as it was very dark at the time.   He afterwards learned that the same appearance had been observed
by several pilots exposed to the same shower, and he attributed the occurrence to a species of St. Elmo's fire.

CHAPTER XVI.

LUMINOUS FLOWERS.

AMONG the earliest observers of phosphorescent flowers may be mentioned a young Swedish girl, the daughter of Linnæus, the eminent naturalist. While walking in the garden one sultry night, she saw what was described as a "lightning-like phosphorescence" about the flowers of the nasturtium (*Tropæolum majus*). The sparks, or flashes, were also visible early in the morning, but, curiously enough, were not apparent in complete darkness; the time between day and night evidently being the most favorable for the exhibition. This observation was made in 1762, and the young girl lived to the advanced age of ninety-six, often repeating the story.

In 1843 Mr. Dowden, an English botanist, noted a similar display in the double variety of a common marigold. Several friends were with him at the time; and, by shading the flower, they distinctly saw a golden-colored lambent light playing from petal to petal, so that an almost uninterrupted corona was formed about the disk.

Others have observed this peculiarity in this flower and in the hairy red poppy (*Papaver pilosum*). A correspondent of the "Gardner's Chronicle" writes, "We witnessed (June 10, 1858) this evening, a little before nine o'clock,

a very curious phenomenon. There are three scarlet verbenas, each about nine inches high, and about a foot apart,
planted in front of the greenhouse. As I was standing a
few yards from them, my attention was arrested by faint
flashes of light passing backwards and forwards from one
plant to the other. I immediately called the gardener and
several members of my family, who all witnessed the extraordinary sight, which lasted for about a quarter of an hour,
gradually becoming fainter, till at last it ceased altogether.
There was a smoky appearance after each flash, which we all
particularly remarked. The ground under the plants was
very dry ; the air was sultry, and seemed charged with electricity. The flashes had the exact appearance of summer
lightning in miniature. This was the first time I had ever
seen any thing of the kind ; and having never heard of such
appearances, I could hardly believe my eyes. Afterwards,
however, when the day had been hot and the ground was
dry, the same phenomenon was constantly observed at about
sunset, and equally on the scarlet geraniums and verbenas.
In 1859 it was again seen. On Sunday evening, June 10,
of that year, my children came running in to say that the
lightning was again playing on the flowers. We all saw it;
and again on July 11, I thought that the flashes of light
were brighter than I had ever seen them before."

It has been asserted that this phenomenon was due to
optical illusion, but the experience of Goethe points to a different conclusion. He says, "On the 19th of June, 1799,
late in the evening, when the twilight was passing into a
clear night, as I was walking up and down with a friend in
the garden, we remarked very plainly about the flowers of
the Oriental poppy, which were distinguishable above every

thing else by their brilliant red, something like flame. We placed ourselves before the plant, and looked steadfastly at it, but could not see the flash again, till we chanced in passing and repassing to look at it obliquely; and we could then repeat the phenomenon at pleasure. It appeared to be an optical illusion, and that the apparent flash of light was merely the spectral representation of the blossoms of a blue-green."

It is an interesting fact, that the light has been observed principally about yellow flowers, as the sunflower (*Helianthus annuus*), the *Rose d'Inde* and *Oeillet d'Inde*, the garden marigolds (*Calendula*), yellow lily, and others.

The Swedish naturalist, Professor Haggern, was fortunate in observing the light about the marigold. His first impression was, that it was an illusion; and to convince himself, he placed a man near at hand with orders to make a signal when he saw the light: the result was, that both observed it simultaneously. The light appeared as a flash, often in quick succession from the same flower, and again only after several moments. It was only observed at sunset on dry days. Professor Haggern's observations were made upon the marigold, garden nasturtium (*Tropæolum majus*), the orange lily (*Lilium bulbiferum*), and the French and African marigolds (*Tagetes patula* and *T. erecta*). He was at first disposed to consider the light due to some insect, but finally decided that it was electrical.

In 1857 the press of Upsala, Sweden, contained accounts of remarkable lights that had been observed about a group of poppies in the Botanic Gardens. The observer, M. Th. Fries, a well-known botanist, in passing the flowers, noticed three or four emitting little flashes of light. Believing that he was the victim of an optical illusion, and wishing to satisfy himself, he took a friend to the place at the same hour on the following night, without, however, informing him what he had seen. The latter immediately noticed the light, and soon the garden was thronged with persons interested, who wished to see the flowers that "threw out flames." Later, fourteen persons saw the exhibition at once, not only upon the *Papaver orientale*, but on the *Lilium bulbiferum;* and before the curious phenomenon ceased, over one hundred and fifty reliable observers were enabled to testify to the delight they had experienced in watching the gleams of light play about these flowers; the doubters and critics, of which there were many, being effectually silenced.

It is usually the misfortune of the single observer, or the minority, to be ridiculed, and their word doubted, simply because others do not choose to believe their statements. That such a course is unjust, is well shown in the instance of the daughter of Linnæus, who made the statement, that as she approached the flowers of *Dictamnus albus* with a light they appeared to ignite, without, however, injury to them. This experiment was tried time and again by others, but without success; and not a few scientists of the day regarded it as a delusion, while others averred that it was pure invention; opinions which placed the lady in a disagreeable position. Some years after, Dr. Hahn was enabled to show that the experiment was not mere fiction. He says, "Being in the habit of visiting a garden in which strong, healthy plants of *Dictamnus albus* were cultivated, I often repeated the experiment, but always without success; and I already began to doubt the correctness of the observation made by the daughter of Linnæus, when, during the dry and hot summer

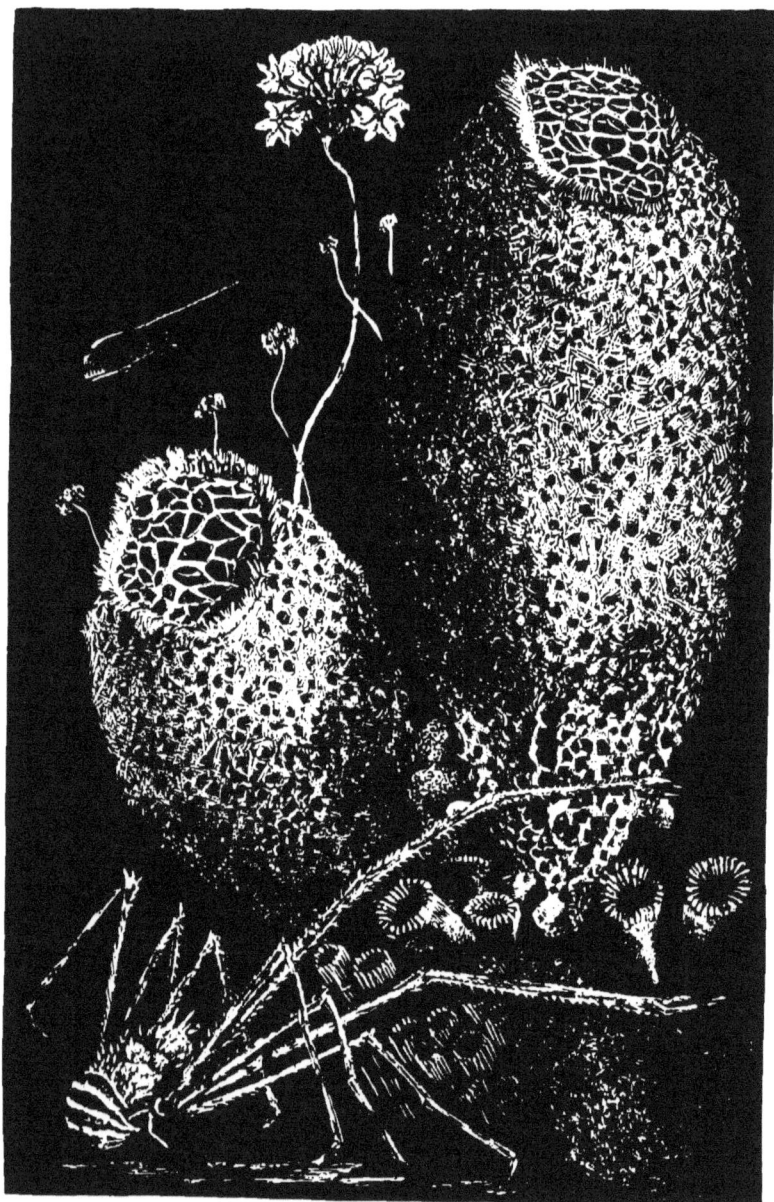

LUMINOUS UMBELLULARIA.

LUMINOUS FISH.   SILICIOUS SPONGE.
(*Chauliodus.*)

                                       LUMINOUS CORALS.
LUMINOUS CRUSTACEAN.
(*Ptichogaster.*)

of 1857, I repeated the experiment once more. Fancying that the warm weather might possibly have exercised a more than ordinary effect upon the plant, I held a lighted match close to an open flower, but again without result; in bringing, however, the match close to some other blossoms, it approached a nearly faded one, and suddenly was seen a reddish, crackling, strongly shooting flame, which left a powerful aromatic smell, and did not injure the peduncle. Since then I have repeated the experiment during several seasons; and even during wet, cold summers it has always succeeded, thus clearly proving that it is not influenced by the state of the weather. In doing so, I observed the following results, which fully explain the phenomenon. On the pedicels and peduncles are a number of minute reddish-brown glands, secreting etheric oil. These glands are but little developed when the flowers begin to open, and they are fully grown shortly after the blossoms begin to fade, shrivelling up when the fruit begins to form. For this reason the experiment can succeed only at that limited period when the flowers are fading. The radius is uninjured, being too green to take fire, and because the flame runs along almost as quick as lightning, becoming extinguished at the top, and diffusing a powerful incense-like smell."

As to the actual cause of these exhibitions of light, little is known. In the case of M. Fries, the luminosity was always observed between quarter past ten and quarter past eleven in the evening, and especially when the weather was sultry, and was seen to best advantage when the observers were not looking at the flowers directly. The light appeared in fitful flashes, similar to that seen about the other flowers mentioned, and was supposed to be electric. Mr. Haggern was also inclined to believe that the phenomenon observed by him was electric. The flame seen about the flower ignited, by the daughter of Linnæus, was caused, as suggested by Phipson, by the ignition of the inflammable atmosphere that envelopes the essential oil-glands of certain *Flaxinellæ.* Electric light has been observed in a plant allied to the palm, belonging to the genus *Pandanus.* When the spatha, or covering which envelopes the flowers, is ruptured, a crackling sound is heard, and a spark of light emitted. It is not impossible that the light of certain flowers is in some way attendant upon the escape of pollen.

CHAPTER XVII

ONE of the most remarkable and awe-inspiring phenomena of the ocean is the water-spout, — a lofty column
composed of tons of water, whirling upward, lifted by the
mighty force of the wind. From a distance the formation
of a spout is an interesting sight. In my own observations,
there has generally been a low-lying bank of dark lead-col-
ored cloud to announce its coming. From this a sharp cone,
seemingly of cloud, was seen to drop, and in a very few
moments an attenuated pillar rose from the water directly
beneath it. The two appeared to meet, and, the alliance
consummated, the lofty column moved away with a greater
or less velocity.

Near proximity to them is not unaccompanied with dan-
ger; and I once found myself in the centre of four or
five, which were moving slowly about. The wind almost
entirely died away, so that had our boat been a large one,
we would have been completely at the mercy of the aqueous
giants; as it was, we lowered the sail, and taking the oars,
succeeded in avoiding them all.

It is the general impression, that if a water-spout touches
an object, or is struck, its form is broken, and the water
descends; but this is not always the case. I was stand-

ing one day upon the sea-wall of Fort Jefferson, on the
island of Tortugas, Florida reef, when I perceived a lofty
water-spout, a mile to the east, headed directly for the fort,
as I thought. In a few moments it struck Long Key, a
narrow island a quarter of a mile away; passing over perhaps
one hundred and fifty feet of it, striking a small schooner
which had been hauled upon the beach, twisting it around,
and then continuing its course with great rapidity. It now
turned a little to the north; and, seeing that in all probability
it would not strike the fort, I awaited its coming. Never
shall I forget the awful grandeur of the sight, as the watery
monster, seemingly several thousand feet in height, reached
the shoal. For some reason which I cannot explain, the
central portion was invisible, but the upper part was distinctly seen, and appeared to be nearly over my head; and
its proximity may be imagined from the fact that the drops
from it seemed like a heavy rain. The entire spout was
bent like a bow by the wind, and was moving along with
great rapidity. I could not keep up with it, though running at utmost speed as it passed. Its progress was accompanied by a loud roar, and a hissing, splashing sound,
while great masses of foam were thrown up before and
behind. In its wake followed numbers of gulls, feeding
upon the small fishes killed by the rush of waters; and
where it crossed the shoal, in perhaps eight feet of water,
quite a trench was scooped out. Imagine such a column at
night coursing over the ocean; its entire shape outlined
against the darkness in phosphorescent light (Plate XXV.),
and an idea may be gained of the magnificent spectacles
which, on rare occasions, are produced by some of the simplest of plants, — the diatoms,[45] whose nuclei are luminous.

The southern oceans, in certain places, often swarm with these minute light-givers, and when borne aloft in the spout, they tend to produce one of the most remarkable and striking scenes possible to imagine. In color these luminous columns are yellow, of different shades, according to the numbers of diatoms present. The naturalists of the "Challenger" found that *P. pseudo-noctiluca* was always present, and often existed at the surface in vast numbers, in the tropics and subtropical regions where the temperature was over sixty-eight or seventy degrees; and the most beautiful exhibitions seen during the cruise were due to these little forms. They have been observed in the Bay of Funchal all the year round. The light was equally brilliant in each species; and in each, when disturbed several times in succession, the phosphorescence perceptibly diminished, and finally disappeared; but after an hour's rest, it re-appeared as brilliant as before.

The phosphorescence of plants, though not so remarkable in its general manifestations as in the forms previously reviewed, is sufficiently interesting to attract general attention. In nearly all countries these vegetable lamps are found; and even in the old legends of the Greeks, Hindus, and Persians, references to the "burning bush," and other luminous phenomena are met with, evidently having some foundation in fact. In India the old natives tell the story, that their forefathers, who visited the mountain of Sufed Koh, at the north of Nalroo in Afghanistan, found a spring in which grew a bush which, from a distance, seemed to emit a brilliant light; but if any one approached, it immediately disappeared, vanishing in the air. In 1845 the white residents of Simla were informed by the natives that a wonderful plant was illuminating the mountains near Syree; and those who investigated it expressed the belief that the light, if it existed at all, came from a species of *Dictamnus*, which was known to grow about Gungotree and Jumnotree.

Even in Josephus we find reference to the luminosity of plants. "There is a certain place," he says, "called Baaras, which produces a root of the same name with itself; its color is like to that of flame, and towards evening it sends out a certain ray like lightning; it is not easily taken by such as would do it, but recedes from their hands."

In the "Proceedings of the Royal Asiatic Society" of April, 1845, there is reference to a luminous root-stock found in the Oraghum jungles, "gleaming in the dark with all the vividness of a glow-worm, or the electric scolopendra, after having been moistened with a wet cloth applied to its surface for an hour or two, and did not seem to lose the property by use, becoming lustreless when dry, and lighting up again whenever moistened." It is probable that this is the plant which is referred to by the Brahmins as *Jyotismati*, produced, it is said, by a variety of *Cardiospermum*. According to Sanscrit authorities, it abounds in the Himalaya Mountains; and is well known, according to Major Madden, at Almora, where investigation showed it to be, at least in this locality, the roots of the fragrant khus-khus grass, which at certain times, as rainy nights, was luminous.

In South America and Asia occurs a plant known to science as *Euphorbia phosphorea;* which emits, when severed or cut, a milky juice somewhat resembling that of the dandelion. At night the juice of the former is, when heated, brilliantly phosphorescent; so much so, that, according to M. Martins of Montpellier, if the stem be broken and used as a pen, this *latex* may be employed as a luminous ink, the characters appearing in the dark as letters of fire. One of the most familiar exhibitions of vegetable luminosity is seen in the "touchwood" or "fox-fire," which many a school-boy has employed in the perpetuation of a practical joke. It is found about old decayed trees, and is simply rotten wood permeated by the mycelium of fungi, which is luminous in the dark. This simple luminant is often quite sufficient to enable one to read large print, and is often the cause of laughable episodes among camping-parties. A friend of the writer, in building a camp-fire in the deep woods, hauled an old log to the door of the tent, and there broke it up, making a fire about which the men slept. In the night, after the fire was extinguished, one of the party awoke, and with a shout aroused the rest, who sprang to their feet, believing that they were lying among coals; as all about were masses of wood seemingly at a white heat, but which investigation showed to be fox-fire.

This luminous decayed wood often rolls out from trees in the forests, to the astonishment or alarm of animals un-familiar with fire.

Perhaps the most remarkable exhibition of fox-fire is re-corded by the Rev. M. J. Berkeley, who says, "A quantity of wood had been purchased in a neighboring parish, which was dragged up a very steep hill to its destination. Amongst them was a log of larch or spruce, it is not quite certain which, twenty-four feet long, and a foot in diameter. Some young friends happened to pass up the hill at night, and were surprised to find the road scattered with luminous patches, which, when more closely examined, proved to be portions of bark or little fragments of wood. Following the track, they came to a blaze of white light which was perfectly surprising; on examination it appeared that the whole of the inside of the bark of the log was covered with a white byssoid mycelium of a peculiarly strong smell, but unfortunately in such a state that the perfect form could not be ascertained. This was luminous; but the light was by no means so bright as in those parts of the wood where the spawn had penetrated more deeply, and where it was so intense that the roughest treatment scarcely seemed to check it. If any attempt was made to rub off the luminous matter, it only shone the more brightly; and when wrapped up in five folds of paper the light penetrated through all the folds on either side as brightly as if the specimen was exposed; when, again, the specimens were placed in the pocket, the pocket when opened was a mass of light. The luminosity had now been going on for three days. Unfortunately we did not see it ourselves till the third day, when it had, possibly from a change in the state of electricity, been somewhat impaired; but it was still most interesting, and we have merely recorded what we saw ourselves. It was almost possible to read the time on the face of a watch, even in its less luminous condition. We do not for a moment suppose that the mycelium is essentially luminous, but are rather inclined to believe that a peculiar occurrence of climatic conditions is necessary for the production of the phenomenon, which is certainly one of great rarity. Observers as we have been of fungi in their native haunts for fifty years, it has never fallen to our lot to witness a similar case before; though Professor Churchill Babington once sent us specimens of luminous wood, which had, however, lost their luminosity before they arrived. It should be observed that

LUMINOUS FISH.

(*Malacosteus niger.*)

With two luminous disks, one emitting a golden, the other a greenish light.

the parts of the wood which were most luminous were not only deeply penetrated by the more delicate parts of the mycelium, but were those which were most decomposed. It is probable, therefore, that this fact is an element in the case as well as the presence of fungoid matter."

Any one who has wandered among old tree-trunks in search of insects, or been a careful observer in underground nooks and corners, must have seen the white tangles, often of beautiful shape, which constitute the forms of some fungi. They are frequently to be seen under old boards in frost-like designs of great delicacy, and many of these are supposed by some to have a certain relation to luminous woods. Around old tree-stumps, the decayed arms of the oak especially, long, cylindrical, flexible branches with a hard bark covering are often found. When freshly broken, the interior is pure white, later changing to a more or less deep brown tint. The white, flocculent extremities form the mycelium of the fungus known as *Rhizomorpha subterranea,* one of the most interesting of the luminous plants. Its mystic light is often seen in caves, where the rootlets have made their way, gleaming with a soft phosphorescence.

In coal-mines this plant is quite common, and has been especially observed near Dresden. Ehrman speaks in enthusiastic terms of these " vegetable glow-worms," as he calls them, which he observed gleaming on the walls and in the crevices of Swedish mines.

In Bohemia the caves are not uncommonly illumined by this interesting cryptogam ; and, according to Phipson, sufficient light has been emitted in English coal-mines from this source to enable miners to read ordinary print. In the mines of North Hesse, Germany, the conditions are particularly favorable for such displays, the gleams being described as resembling moonbeams stealing through the gloomy caverns.

That this fungus is luminous when detached, is shown by the following from M. Tulasue, in the "Annals of Natural Science," 1848. "On the evening of the day I received the specimens," he writes, "the temperature being about 22° C., all the young branches brightened with an uniform phosphoric light the whole of their length. It was the same with the surface of some of the older branches, the greater number of which were still brilliant in some parts, and only on their surface. I split and lacerated many of these twigs, but their internal substance remained dull. The next evening, on the contrary, this substance, having been exposed to contact with the air, exhibited at its surface the same brightness as the bark of the branches. Prolonged friction of the luminous surfaces reduced the brightness, and dried them to a certain degree, but did not leave on the fingers any phosphorescent matter." And again, "By preserving these *Rhizomorphæ* in an adequate state of humidity, I have been able for many evenings to renew the examination of their phosphorescence; the commencement of desiccation, long before they really perish, deprives them of the faculty of giving light."

Rumphius, the celebrated botanist, was perhaps the first European to discover the phosphorescence of fungi, observing it in a large specimen on the island of Amboine, which he named *Fungus igneus*, or fire-mushroom. In America such exhibitions are rare. Mr. H. K. Morrell, editor of "The Gardiner (Me.) Home Journal," informed me some few years ago that he had observed the phosphorescence of

Tíanus stypticus in his garden; the young of which, being especially brilliant, emitted a steady light. In Brazil a certain agaric is famous for its vivid luminosity. It was observed by Mr. Gardner in 1840, who says, referring to the species which has been named *Agaricus gardneri,* " One dark night about the beginning of December, while passing along the streets of the Villa de Natividate, Goyaz, Brazil, I observed some boys amusing themselves with some luminous object, which I at first supposed to be a kind of large fire-fly; but, on making inquiry, I found it to be a beautiful phosphorescent species of *Agaricus,* and was told that it grew abundantly in the neighborhood on the decaying fronds of a dwarf palm. The whole plant gives out at night a bright phosphorescent light, somewhat similar to that emitted by the larger fire-flies, having a pale greenish hue. From this circumstance, and from growing on a palm, it is called by the inhabitants ' Flor de Coco.' "

Dr. Cuthbert Collingwood had a similar experience with an allied species in Borneo. " The night being dark, the fungi could be very distinctly seen, though not at any great distance, shining with a soft, pale greenish light. Here and there spots of much more intense light were visible, and these proved to be very young and minute specimens. The older specimens may more properly be described as possessing a greenish, luminous glow like the glow of the electric discharge; which, however, was quite sufficient to define its shape, and when closely examined, the chief details of its form and appearance. The luminosity did not impart itself to the hand, and did not appear to be affected by the separation from the root on which it grew, at least not for some hours. I think it probable that the mycelium of this fungus is also luminous; for, upon turning up the ground in search of small, luminous worms, minute spots of light were observed, which could not be referred to any particular object or body, when brought to the light and examined, and were probably due to some minute portions of its mycelium." Mr. Hugh Low has stated that "he saw the jungle all in a blaze of light, by which he could see to read, as some years ago he was riding across the island by the jungle road, and that this luminosity was produced by an agaric."

Australia has produced a number of luminous toadstools. Drummond found some striking forms near Swan River. He had noticed two species growing as parasites on the stumps of trees. Their appearance in the daytime did not attract particular attention; but at night they developed into veritable plant lamps, exceeding any thing that he had ever seen. One was about two inches across, and grew in clusters on the stump of a banksia-tree which was surrounded by water. When the little plant was secured from its miniature island home, it could have been used as a lamp for several successive nights, a newspaper being read by placing the agaric on it, the light illuminating the type in the immediate vicinity. As the plant dried, the light gradually diminished.

Later Mr. Drummond found a giant specimen that was sixteen inches in diameter and a foot high, a veritable chandelier. He says regarding it, "This specimen was hung up inside the chimney of our sitting-room to dry; and, on passing through the apartment in the dark, I observed the fungus giving out a most remarkable light, similar to that described above. No light is so white as this, at least none

LUMINOUS MUSHROOMS.

LUMINOUS INSECT.
(*Geophilus electricus.*)

that I have ever seen. The luminous property continued, though gradually diminishing, for four or five nights, when it ceased on the plant becoming dry. We called some of the natives, and showed them this fungus when emitting light. The room was dark, for the fire was very low and the candles extinguished; and the poor creatures cried out, 'Chinga,' their name for a spirit, and seemed afraid of it."

A very attractive agaric, *Agaricus olearius* (Plate XIII.), is found at the foot of olive-trees in Southern Europe. During the daytime the color is yellow, but observed at night it emits a brilliant blue light. Like the Australian species, it continues to emit light after it has been taken from the ground, the phosphorescence persisting for successive nights. So brilliant are the gleams, that they may be perceived at times before darkness sets in. Experiment showed that the light was extinguished when the temperature was below + 90° to + 6° C.; but the luminosity was not destroyed, as it re-appeared when the temperature was raised above this point. If kept some time in a temperature below freezing, it loses its light-emitting property entirely. It gleams as brightly under water as out; pure oxygen seems to have no effect upon it, and the most careful experiments fail to show the slightest elevation of temperature about the parts which shine. The light seems to emanate from the head (*pileus*) of the fungus, the *lamellæ* of the latter, where the seeds are found, being the centre of the luminous phenomenon.

These interesting light-givers are perhaps more common than we are aware of, from the fact that nocturnal investigations in the woods are not frequent, nearly all the discoveries being the result of accident. A small, luminous fungus has been observed in the Andaman Islands. Gandichand found one in Manilla, while Dr. Hooker, as we have seen, refers to the presence of one in the Sikkim Himalayas.

These curious families of fungi are not only ornamental, but useful. In European countries the common mushroom [46] enjoys the widest popularity as an esculent, especially the cultivated varieties. The meadow mushroom is scarcely inferior, though stronger in flavor, and is preferred by many to the cultivated species. In France the champignon is largely eaten; and in Austria a kind which has no admirers in England finds a constant place in the markets during the summer. Truffles and morels are favorites not only in Europe, but also in the vales of Cashmere, where two or three species of morels are dried for consumption throughout the year. The great puff-ball is increasing in reputation as a breakfast delicacy in Great Britain, while the chantarelle and the hedgehog fungus are esteemed by many.

Numerous other species are more or less eaten by mycophagists, although they are never found in the public markets. A species of *Boletus*, cut in slices and dried, may be purchased throughout the year in most of the Continental cities. In Tahiti the Jew's ear [47] is dried in large quantities and exported to China; while a species of agaric comes into the markets of Singapore, and another dried agaric is sent from the Cabul hills and the plains of north-western India. Several species of *Cyttaria* are eaten in the southern parts of South America, and in Australia a native kind [48] is a favorite article of food. Indeed, a very long catalogue might be made of the species which are more or less consumed in different parts of the world.

The cultivation of fungi for esculent purposes has not hitherto been successful with any other species than the ordinary mushroom. Attempts were made in France to cultivate truffles, at first apparently with considerable promise, but ultimately without much satisfaction. There is no good reason to suppose it impossible or improbable that many species might be devoted to experiments in that direction. Some species of *Pólyporus* have been employed as styptics, or beaten till soft and used as amadou. One species in Burmah has a good reputation as an anthelmintic. Some species of *Polysaccum* and *Geaster* are employed medicinally in China. Species of *Elaphomyces* were at one time supposed to possess great virtues now deemed apocryphal. Ergot, developed on rye, wheat, and the germen of various grasses, still maintains its position in the pharmacopœia; but is almost the only fungus now employed, and that sparingly, by the legitimate medical practitioner.

In the Cardiff coal-mines an interesting plant is found, which emits so brilliant a light, that the men have been able to "see their hands by it," and was visible at a distance of sixty feet. Mr. Worthington Smith, who is authority for this, observed the same phenomenon in *Polyporus sulfureus*.

While various theories have been recorded as to the physiological cause of the light in cryptogams, and many writers give the most careful details of the structure of the luminous parts, we are unable to go a step farther to explain the cause of the light which appears to be a combustion, but does not consume.

CHAPTER XVIII.

PHANTOMS.

PHOSPHORESCENT light plays an important part in the composition of ghosts and phantoms; and the number of persons who believe that certain phenomena exist which cannot be explained by well-known natural laws is somewhat surprising. Some years ago I was introduced to a gentleman who was a firm believer in a modern Flying Dutchman. His house was upon a beautiful little bay, and from the piazza, he informed me that, more than once, he had seen a phantom ship. Sometimes it beat up the bay, the white sails showing distinctly at night. Again it was seen coming in directly against the wind, now appearing in one place, then in another, as fickle as the wind itself. On every other subject he was sane, and of more than ordinary intelligence; but some electric phenomenon or emanations from schools of fishes, together with a vivid imagination, had produced the phantom ship, which in his mind was a reality.

Many well remember the excitement occasioned around one of the New-York markets a number of years ago, by the appearance of a mysterious light. A fish-dealer's assistant, who had occasion to enter the market late one evening, observed an unusual light there; and being an ignorant, superstitious fellow, he rushed out of the building and into a neighboring store, stating excitedly that the ghost of a former market-man was hovering about his old stand. A number of persons returned with him to the market, and there saw a light, a dull yellowish gleam, about six feet in length, proceeding apparently from some body lying in a recumbent position. The crowd pressed in, and found the ghost to be a large piece of fish that had become phosphorescent.

Such occurrences are not uncommon, and show that phosphorescence is not confined to any special place, object, or condition. As early as 1592 we read of its having caused surprise and astonishment among the Romans. Several young men having bought a lamb, and kept it over night for an Easter feast on the following day, were amazed to find that at night the flesh gleamed as if candles had been placed upon it. So much interest was aroused by the occurrence, that the animal was sent to a scientist of the day, Fabricio d'Acquapendente, for explanation; but it was as little understood then as it is to-day. This meat emitted a white light, and it was communicated to a piece of kid's meat that was placed in contact with it.

Bartholin, the Danish philosopher, records an instance that excited much interest in his day. A poor woman had purchased a piece of meat; and, during the night having to go to the pantry, was terrified by observing that it was surrounded by a blaze of light. Many persons visited the house, and it was noticed that as soon as putrefaction commenced the light disappeared.

According to M. Nueesh, in a certain butcher's shop the meat became strongly phosphorescent, and remained so as long as sound. If putrefaction set in, and *Bacterium termo* made its appearance, the luminous appearance ceased. In many cases timid persons have thrown water upon such light, but without effect. Alcohol and certain acids, however, seem to extinguish it. Boyle was curious enough to place a piece of shining veal in the receiver of an air-pump, which had no perceptible effect upon it, showing that there was no combustion, as we understand it. He also used his luminous meat as a lamp, and states that it made a "splendid show." A printed paper was placed over the light spots, and the type made out without difficulty.

If heat is given out by this light, the instruments of the present day fail to show it. Every surgeon has had experience with this phenomenon in the course of his studies, yet it is still unexplained.

We have observed living forms producing light from special plates, or from the mucilaginous envelope of their bodies, and when dead the same curious light appears for a limited time. Dr. Phipson examined a luminous ray with great care, thinking to find traces of phosphorus in the luminous grease, but it was entirely wanting. The little boring-shell pholas, which we have seen is a brilliant light-giver when alive, is equally so after death; its luminosity continuing in honey for a year, as previously described.

A boat containing dead mackerel often presents the appearance of being loaded with coals of fire, each fish gleaming with a soft phosphorescent light, that seems to arise in the greasy mucus which covers them. Place one of these luminous fishes in the water, and the latter will soon assume a like appearance. Vegetables piled in cellars often appear phosphorescent, especially potatoes and cantelopes. In a case of the former, a servant seeing the brilliant light gave an alarm of fire, arousing the neighborhood. The men rushed in, and the cellar was well flooded before it was discovered that some unoffending potatoes were the cause of the alarm.

CHAPTER XIX.

LUMINOUS SHOWERS.

IN many old works, accounts are found of so-called show-ers of fire, during which the entire heavens seemed filled with gleaming drops, that threatened to burn every living thing, but were in reality harmless; the exhibition being merely another instance of this strange phenomenon of heatless light.

Some years ago a party of peasants were making the ascent of one of the high peaks of the Alps, when they were caught in a rain-storm, which produced a demoralizing effect upon them. As the rain fell, it seemed to become luminous, and drops of fire apparently ran from their cloth-ing and beards. Their attempts at brushing it away, while adding to the startling nature of the phenomenon, showed, however, that it was perfectly harmless.

Dr. Phipson records some interesting instances of this kind of phosphorescence, of which the following may be cited: —

M. de Thielan observed on Jan. 25, 1822, near Freyburg, a most extraordinary spectacle. A heavy snow had been falling during the early part of the evening, and the trees, branches, limbs, and leaves quivered and scintillated with a resplendent bluish light, while the drops of rain upon the grass left golden trains as they dripped to the ground.

DEEP SEA ANGLER.
(*Melanocetus johnstoni, Gth.*)

Arrago records similar occurrences: In 1731 a priest named Hallai, who lived at Lessay, near Constance, states that he observed one evening during a severe thunder-storm, rain falling which looked like *drops of red-hot liquid metal.*

Bergman, the eminent Swedish chemist, communicated to the Royal Society of London, in 1761, that late in the afternoon upon two occasions, though hearing no thunder, he had seen rain which glittered as it fell upon the ground, making it look as if covered with waves of fire.

M. Pasumot, on May 3, 1768, was overtaken, while walking near Arnay-le-Duc, on an open plain, by a very heavy storm. The rain collecting on the brim of his hat, he stooped his head to allow it to run off, when to his astonishment, as it encountered that which fell from the clouds, at about twenty inches from the ground, *sparks were emitted* between the two portions of liquid.

During January, 1822, Lampadius was told by the miners of Freyburg, that they had observed during a storm, *sleet* which *emitted light* when it fell upon the ground.

A friend of Howard, the meteorologist, stated to him, that while going from London to Bow on the 19th of May, 1809, there came up a very severe storm; and he observed the rain emit light as it struck the earth.

On the 28th of October, 1772, the Abbé Bertholon, who was travelling to Lyons from Brignai, early in the morning was overtaken by a violent storm of rain and hail. The *rain and hail-stones emitted light* as they fell upon the metallic mounting of his horse's trappings.

Luminous hail has often been observed; and when we remember that hail-stones sometimes attain great size, we can imagine the scene occasioned by a fall where each stone is phosphorescent. Ordinary hail-stones are the size of small peas, but they occasionally occur large enough to kill human beings; and I have seen them so large in the Sierra Madre Mountains that any shelter was preferable to exposure to them. In 1707 a hail-storm occurred at the town of Como, Switzerland, doing an incredible amount of damage, some of the stones weighing nearly ten ounces. Darwin describes a storm upon the South American pampas, in which the stones that fell were large enough to kill powerful animals.

Ice has often been observed to emit luminous sparks; and probably one of the grandest spectacles ever witnessed, is the luminous cap of a snow-covered mountain. The glaciers of the Alps have been seen bathed in a soft phosphoric glow, the icy rivers being distinctly marked by the phenomenon, which is so brilliant, at times, that the appearance of a second sunset is occasioned. Not only are the summits of Alpine peaks and the glaciers luminous, but the valleys of Piedmont, Valais, and others have been seen to emit from their covering of snow a soft blue light of singular beauty. So intense is this light about the cap of Mount Blanc, it has been photographed. Luminous vapors or mists may be mentioned in this connection. Several times in the history of this country, luminous mists or fogs have been recorded. Massachusetts was visited by one some years ago, in which the fog was so dense that observers a few feet away were invisible, yet darkness was not an accompaniment; the mist seemed to be light-emitting itself, having a reddish, metallic hue. Others described it as a fiery red or yellow, while to some it appeared to be composed of faintly luminous matter.

In the year 1783 all Europe and a portion of Asia were

enveloped in a dense fog of a most remarkable nature. It was termed "dry," as even at night no dampness was observed. It was first seen at Copenhagen, its coming being heralded by severe storms. A few weeks later it appeared in various parts of France, and rapidly seemed to spread over Europe and portions of Asia. During the day it had a metallic glow, which at night changed to a phosphorescent light, so brilliant that ordinary print could be read by it. Many attempts to explain it were made by the savants of the day, and it was universally supposed to be due to the earthquakes and volcanic eruptions which were of unusual severity that year.

A somewhat similar fog appeared in the United States, a portion of Europe, and Africa in August, 1831. The daylight was perceptibly diminished, while at night a conspicuous phosphorescent light was emitted. A remarkable luminous fog occurred in Switzerland in 1859. M. L. F. Wartmann of Geneva states that the strange light was observed on five successive nights, and apparently proceeded from a heavy dry fog that hung over Geneva during the time. The light was so brilliant that this gentleman distinguished the smallest objects upon his table with perfect ease, no other light being in the apartment. The light caused general comment in the places in which it appeared, and a traveller between Geneva and Annemassi stated that he readily found the road by its means. Dr. Verdeil of Lausanne describes a fog which diffused so much light that distant objects were perfectly visible at night.

Among the phenomena which attended the eruption of Vesuvius in 1794 was one which did not tend to allay the fears of the people. During the day a fine dust filled the air about Naples, which was not particularly noticeable; but as night came on, it emitted a pale though distinct phosphorescent light. An English gentleman sailing near Torre del Greco noticed that where the dust collected upon his hat it was luminous, and no little consternation was caused among the superstitious sailors by the occurrence.

Luminous dust-showers have been noticed in several localities; and the peculiar glows that were seen in this country a few years ago were accredited by many to them, the supposition being that dust, perhaps from volcanic eruptions, was floating about in the upper strata of the atmosphere. Many other explanations were given, and the literature upon the subject is extremely voluminous and interesting.

The amount of material floating about in the upper regions of the air is perhaps little realized by my young readers, and some reference to the phenomenon may be of interest.

Professor Nordenskjöld has for many years been a close observer of dust of all kinds that has fallen upon the earth in rain or snow; and it was his good fortune, during the expedition of the "Vega," to prove beyond a doubt the presence of cosmic dust. For many years we have been assured by astronomers that the earth was being bombarded, as it were, continually, by innumerable meteors. The moment they enter our domain, we observe the spectacle of their ignition. In a moment they are reduced to ashes, and the fine impalpable dust drops slowly, an invisible shower, upon the earth. When such showers are intensified, it is not impossible that some outward and visible phenomena may be the result.

In the search for this cosmic dust, the far North, where the surface is covered by an almost continuous coating of snow and ice, offers a wide and promising field for investigation. Here no other dust prevails. Professor Nordenskjöld first found cosmic dust in the North at Spitzbergen. The second discovery, off the Taimar coast, seemed to be in the form of yellow specks lying on the snow. They were at first supposed to be diatomaceous ooze;[49] but when placed in the hands of Dr. Kjellman, he pronounced them to be pale yellow crystals, and, curious enough, formed of carbonate of lime. " The original composition and origin of this substance," says Professor Nordenskjöld, "appears to me exceedingly enigmatical. It was not carbonate of lime, for the crystals were rhomboidal, and did not show the cleavage of calcite. Nor can there be a question of its being arragonite, because this mineral might indeed fall asunder of itself; but in that case the newly formed powder ought to be crystalline. Have the crystals originally been a new hydrated carbonate of lime formed by crystallizing out at a temperature of ten or twenty degrees above the freezing point? In such case they ought not to have been found on the surface of the *snow*, but lower, on the surface of the *ice*. Or have they fallen down from the inter-planetary spaces to the surface of the earth, and before crumbling down have had a composition differing from terrestrial substances, in the same way as various chemical compounds found in recent times in meteoric stones? The occurrence of the crystals in the uppermost layer of snow, and their falling asunder in the air, tell in favor of this view. Unfortunately there is no possibility of settling these questions; but at all events this discovery is a further incitement to those who travel in the high North, to collect with extreme care, from snow-fields lying far from the ordinary routes of communication,

all foreign substances, though apparently of trifling importance."

The investigations of the Swedish naturalist in this field
are of exceeding interest. His first attempt to obtain meteoric dust was at Stockholm, where, in December, 1871, there
was a great fall of snow, the heaviest ever known. On the
last days of the storm, after the atmosphere had been presumably purified of extraneous substances, he collected a
cubic metre of snow, melted it over a fire, and found that
after the water had evaporated a residue of black powder
remained, in which were many grains of metallic iron, that
were attracted by a magnet. In 1872 his brother made a
similar examination of the snow, in a quiet locality near
the remote village of Evois, Finland. The snow upon
being melted also gave the same black powder and metallic iron.

The investigations of Nordenskjöld himself, conducted in
Spitzbergen, as previously mentioned, were the most satisfactory. The observations were made in 80° north latitude,
and 13° to 150° east longitude, in the layer of snow that
covered the ice. An imaginary section was as follows:
(1), new fallen snow; (2), a layer of hardened old snow,
eight millimetres in thickness; (3), a layer of snow, con-
glomerated to a crystalline granular mass; and (4), common
granular hardened snow. Layer three was full of small
black grains, among which were found numerous metallic
particles, that were attracted by the magnet, and found to
contain iron, cobalt, and possibly nickel also.

In his visit to Greenland in 1870, Nordenskjöld found in
the dust that lay on the inland ice, grains of metallic iron
and cobalt. "The main mass," he says, "consisted of a

crystalline, double refracting silicate, drenched through with an ill-smelling organic substance. The dust was found in large quantities at the bottom of innumerable small holes in the surface of the inland ice. This dust could scarcely be of volcanic origin, because by its crystalline structure it differs completely from the glass dust that is commonly thrown out of volcanoes, and is often carried by the wind to very remote regions; as also from the dust which, in March, 1875, fell at many places in the middle of Scandinavia, and which was proved to have been thrown out by volcanoes in Iceland." Professor Nordenskjöld's estimate of the quantity of dust shows that it has been in past ages a not unimportant factor, perhaps, in its addition to the crust. He says, " I estimate the quantity of the dust that was found on the ice north of Spitzbergen, at from .01 to 1 milligram per square metre ; and probably the whole fall of dust for the year far exceeded the latter figure. · But a milligram on every square metre of the surface of the earth amounts for the globe to five hundred million kilograms (say half a million tons). Such a mass, collected year by year during the geological ages, of a duration probably incomprehensible by us, becomes a consideration too important to be neglected, when the fundamental facts of the geological history of our planet are enumerated. A continuation of these investigations will perhaps show that our globe has increased gradually from a small beginning to the dimensions it now possesses ; that a considerable quantity of the constituents of our sedimentary strata, especially of those that have been deposited in the open sea far from land, are of cosmic origin ; and will throw an unexpected light on the origin of the fire-hearths of the volcanoes, and afford a simple explanation of the remarkable resemblance which unmistakably exists between plutonic rocks and meteoric stones."

Such enormous masses of material could well explain the rosy and other curious lights that are from time to time observed. But cosmic dust is not the only matter in the air that could occasion the phenomena; the atmosphere is constantly filled with innumerable forms caught up by currents and carried to inconceivable heights, and thus to great distances, to be precipitated to the earth in hail, snow, or rain.

Near St. Domingo, Darwin tells us, the atmosphere became thick and hazy from the impalpable fine dust that actually injured their astronomical instruments. " The morning before we anchored at Porto Praya," he says, "I collected a little packet of this brown-colored dust, which appeared to have been filtered from the wind by the gauze of the vane at the masthead. Mr. Lyell has also given me four packets of dust which fell on a vessel a few hundred miles northward of these islands. Professor Ehrenberg finds that this dust consists in great part of *infusoria* with siliceous shields, and of the siliceous tissue of plants. In five little packets which I sent him, he has ascertained no less than sixty-seven different organic forms. The *infusoria*, with the exception of two marine species, are all inhabitants of fresh water. I have found no less than fifteen different accounts of dust having fallen on vessels when far out in the Atlantic. From the direction of the wind whenever it has fallen, and from its having always fallen during those months when the harmattan is known to raise clouds of dust high into the atmosphere, we may feel sure that it all comes from Africa. It is, however, a very singular fact, that, although

Professor Ehrenberg knows many species of *infusoria* peculiar to Africa, he finds none of them in the dust which I sent him; on the other hand, he finds in it two species which hitherto he knows as living only in South America; The dust falls in such quantities as to dirty every thing on board, and to hurt people's eyes; vessels even have run ashore, owing to the obscurity of the atmosphere. It has often fallen on ships when more than a thousand miles from the coast of Africa, and at points sixteen hundred miles distant in a north and south direction. In some dust which was collected on a vessel three hundred miles from the land, I found particles of stone, above the thousandth of an inch square, mixed with finer matter. After this fact, one need not be surprised at the diffusion of the far lighter and smaller sporules of cryptogamic plants."

The extent to which dust and ashes can be taken up and held by air currents is shown in volcanoes. In 1810 the ashes from a volcano at St. Vincent were wafted to Barbadoes, nearly a hundred miles; and in 1835 the material thrown from a volcano in Guatamala to Jamaica, eight hundred miles. As intimated, these showers are not all inorganic, but are often living or fossil animals or plants that are floating about. Such are the reddish or gray showers that are frequently met with off the African coast, and when in the snow they are called "blood-rains." The one in 1755, near Lago Maggiore, covered over two hundred square leagues, causing a panic among the inhabitants. For a distance of nine feet below the surface, the snow was blood red, the atmosphere appeared red and fiery, while at sunrise and sunset a rosy hue pervaded every thing. When this shower fell and there was no snow, the earthy deposit accumulated an inch deep; and it has been estimated that, supposing it to average two lines in depth, there would be for each square mile an amount equal to nearly three thousand cubic feet. A similar panic was caused some years ago by a swarm of butterflies. Everywhere they left a drop of blood-colored fluid, so that the fences, houses, and cattle were covered with it. The insects were so numerous that they obstructed the vision.

In the "blood-rains" of Italy, and generally in such instances, the red hue comes from red oxide of iron. At a single shower in Lyons in 1846, Ehrenberg estimated that seven hundred and twenty thousand pounds of material fell, ninety thousand pounds of which were microscopic organisms, including thirty-nine species of siliceous diatoms, and many others of great beauty of form and shape.

Ehrenberg enumerates a very large number of these showers, referring to Homer's "Iliad" for one of the earliest known; and asks, with such facts before us, how many thousand millions of hundred-weight of microscopic organism have reached the earth since Homer's time? The whole number of species made out is over three hundred. The species, as far as ascertained, are not African; fifteen are North American. But the origin of the dust is yet unknown. The zone in which these showers occur covers Southern Europe and Northern Africa, with the adjoining portion of the Atlantic, and the corresponding latitudes in Western and middle Asia.

When blown along by the wind, these showers perform another office besides affecting, perhaps, the color or tint of the atmosphere; they wear away rock, and polish and furrow it. Such work can be seen in the granite rocks

PLATE XXIV.

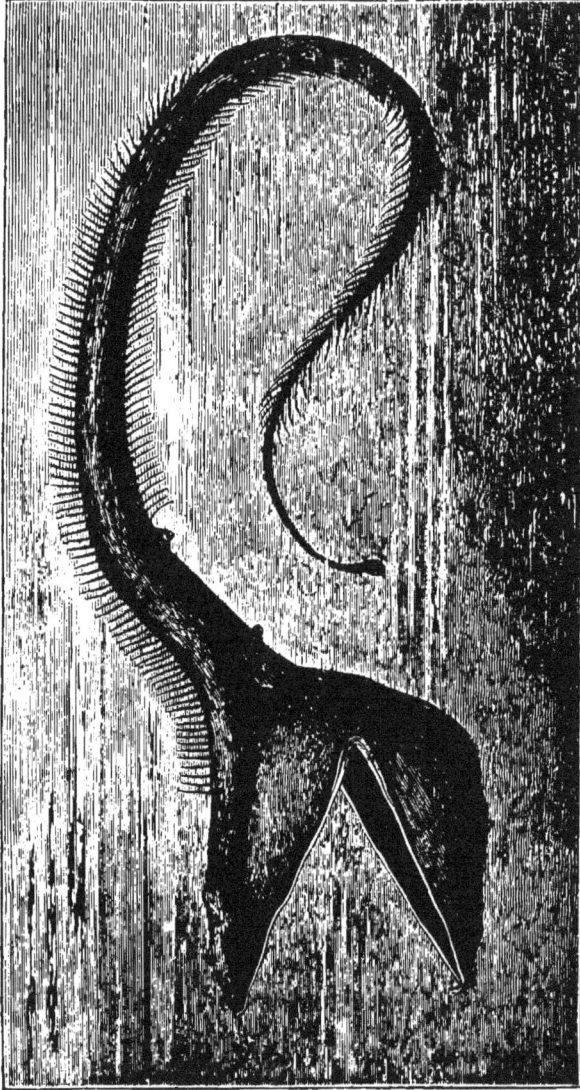

PELICAN FISH.
(*Eurypharyx pelecanoides.*)

at the San Bernardino Pass in California. Quartz is polished, and hard gems left weathered out; while at Cape Cod ordinary sand has been known to wear holes through glass windows by continually blowing against them.

An ingenious instrument has been invented to capture these flying objects of the air. It is called by the inventor, Doctor Miguel, the æroscope, and is really a net for animals invisible to the naked eye.

Many objects are phosphorescent when struck, or when divided into thin laminæ. Some simple materials for such experiments are chlorate of potash, fluor-spar, feld-spar, sugar, etc. By placing any of these in a mortar, and grinding them in the dark, flashes of light will be seen, powdered sugar often making a striking display. A beautiful and effective exhibition can be produced by placing a small amount of phosphuret of calcium in water; decomposition follows, and phosphuretted gas is evolved. As the bubbles of gas rise and come in contact with the atmosphere, they seem to take fire. If in a dark room, luminous rings are seen rising, and they can be made to take various shapes by using a fan. A trick often performed by magicians is to hand around a marble, and then pretend to render it luminous by blowing upon it. This trick consists in having small balls at hand, of a material that can readily be rendered luminous by the application of heat. These substances can be easily made.

A fine light is produced by taking, —

Barium sulphate (C P.) . . . . . . 32 parts
Magnesium carbonate (C P.) . . . . . 1 part
Sulphur (C P.) . . . . 1 part
Gum tragacanth . . . . q. s.

This should be made into balls of a convenient size, dried at a moderate temperature, and kept in a crucible at a red heat for about an hour. Allow them to cool slowly, and then place in a glass-stoppered bottle before their heat has disappeared. When required for use, expose them to the sun or any strong light, and they will become luminous, and continue so for many hours.

Another formula is: —

Strontium sulphate (C P.)	22 parts
Sulphur (C P.)	1 part
Gum tragacanth	q. s.

This should be heated as above described.

A most interesting experiment is to make a selection of artificial flowers, and, having brushed them over with glue or mucilage, dust them with the powder from one of the balls made as described. If the flowers are exposed to the sun a short time, they will emit a phosphorescent light, each flower standing out in the darkness with extreme brilliancy, — a striking and remarkable spectacle.

Canton's phosphorus is easily made by calcining clean oyster-shells, until they are perfectly white, in a crucible. The clearer and finer portions should then be reduced to powder, and placed in layers with alternate layers of flowers of sulphur in a crucible. Cover, and heat to a dull redness for about half an hour, then allow to cool slowly.

Luminous "ink" or liquid can be made by placing a small piece of phosphorus about as large as a pea in a test-tube with a small quantity of olive oil; hold the tube in a water-bath until the oil becomes heated, and the phosphorus liquefies; then shake it until the oil will take up no more phosphorus, and, when it becomes clear, pour into a bottle with a glass stopper. When it is to be used, take out the stopper, and admit the air. The oil can be used with a brush, and in the dark will appear luminous.

Water may be rendered phosphorescent by dissolving a small piece of phosphorus in ether for several days in a glass-stoppered bottle; then by immersing a piece of sugar in the solution, and placing it in water, the latter becomes vividly phosphorescent. It should be remembered that phosphorus and ether are both extremely dangerous, and experiments with them should be conducted with care and judgment.

While this is a mere toy, luminous paint is of great value. It is easily made, and can be applied to many purposes.

Schade of Dresden has quite recently patented an invention, which enables him to produce paints that are luminous without affecting the tint by day. This is accomplished, according to the inventor, as follows: —

Zanzibar or Kauri copal is melted over a charcoal fire. Fifteen parts of the melt are dissolved in 60 parts of French oil of turpentine, and the filtered solution is mixed with 25 parts, previously heated and cooled, pure linseed-oil. The varnish which is thus obtained, is used in the following methods, in the manufacture of luminous paints, by grinding between granite rolls in a paint-mill. Iron rolls should be avoided, because particles of iron, which are liable to be detached, would injure the luminous properties.

Varnishes, as they occur in commerce, generally contain lead or manganese, which would destroy the phosphorescence of calcium sulphide. *A pure white luminous paint* is prepared by mixing 40 parts of the varnish obtained in the above-described process with 6 parts prepared barium sulphate, 6 parts prepared calcium carbonate, 12 parts prepared white zinc sulphide, and 36 parts good luminous calcium sulphide in a proper vessel, to an emulsion, and then grinding it very fine in a color-mill. For *red luminous paint*, 60 parts varnish are mixed with 8 parts prepared barium sulphate, 2 parts prepared madder lake, 6 parts prepared realgar (red arsenic sulphide), and 30 parts luminous calcium sulphide, and treated the same as for white paint. For *orange luminous paint*, 46 parts varnish are mixed with 17.5 parts prepared barium sulphate, 1 part prepared Indian yellow, 1.5 parts prepared madder lake, and 38 parts luminous calcium sulphide. For *yellow luminous paint*, 48 parts varnish are mixed with 10 parts prepared barium sulphate, 8 parts barium chromate, and 34 parts luminous calcium sulphide. For *green luminous paint*, 48 parts varnish are mixed with 10 parts prepared barium sulphate, 8 parts chromium oxide green, and 34 parts luminous calcium sulphide.

A blue luminous paint is prepared from 42 parts varnish, 10.2 parts prepared barium sulphate, 6.4 parts ultramarine blue, 5.4 parts cobalt blue, and 46 parts luminous calcium sulphide.

A violet luminous paint is made from 42 parts varnish, 10.2 parts prepared barium sulphate, 2.8 parts ultramarine violet, 9 parts cobaltous arsenate, and 36 parts luminous calcium sulphide.

For gray luminous paint, 45 parts of the varnish are mixed with 6 parts prepared barium sulphate, 6 parts prepared calcium carbonate, 0.5 parts ultramarine blue, 6.5 parts gray zinc sulphide.

A yellowish-brown luminous paint is obtained from 48 parts varnish, 10 parts precipitated barium sulphate, 8 parts auri pigment, and 34 parts luminous calcium sulphide.

Luminous colors for artists' use are prepared by using East India poppy oil in the same quantity, instead of the varnish, and taking particular pains to grind the materials as fine as possible.

For luminous oil-color paints, equal quantities of pure linseed oil are used in place of the varnish. The linseed oil must be cold-pressed, and thickened by heat. All the above luminous paints can be used in the manufacture of colored papers, etc., if the varnish is altogether omitted, and the dry mixtures are ground to a paste with water.

The luminous paints can also be used as *wax colors for painting on glass* and similar objects, by adding, instead of the varnish, ten per cent more of Japanese wax, and onefourth the quantity of the latter of olive oil. The wax colors prepared in this way may also be used for painting upon porcelain, and are then carefully burned without access of air. Paintings of this kind can also be treated with water-glass.

CHAPTER XX.

THE USES OF PHOSPHORESCENCE.

A S to the value and use of the gift of luminosity possessed by various animals, we can only surmise. Many interesting theories have been suggested, none of which, however, seem to stand the test of practical application. Some naturalists believe that the light of certain invertebrates is a warning. As an example, the jelly-fishes have a terrible array of stings; and it is supposed that fishes once stung, remember the light of these forms, and avoid them in the future. If this were true, many helpless animals, as the salpa and others, would also find protection in the lesson taught by the jelly-fishes.

It is a poor rule that will not work both ways; and we might well ask, if nature supplies these lights as warnings, why the physalia, the most terrible of all these forms, has not been thus provided. Phipson mentions it as a phosphorescent animal, but in the thousands that I have observed during a long residence in the physalia country, I never saw one give out light; hence I assume that if they are luminous, it is only on certain occasions. It might be considered that the vivid colors of this attractive creature constituted a warning; but even this does not hold, as I have found all kinds of pelagic fishes in their toils, and even a turtle and many small fishes bite readily at the deadly tentacles.

It is well known that the sunfish (*Orthagoriscus*), lumpfish, and dogfish all attack jelly-fishes, perhaps in default of better food; and far from being afraid of light, all fishes are attracted by it. It is evident, that, if jelly-fishes possess eyes, they must be able to distinguish others of their kind; hence their phosphorescence may possibly be a simple signal language, if so we may term it, by which they may find one another; or, having its origin in the nervous functions of the animal, the light may be unconsciously emitted, and have no more significance than a blush or sudden pallor upon the human face. Whatever may be the value of the light to themselves, it is of obvious use to other animals. It assists in the general illumination of the deep recesses of the ocean; and, in the case of jelly-fishes, certainly marks their position, and thus aids the whalebone whales when feeding at night at depths from the surface where little light penetrates.

The various colored lights seen upon certain crustaceans and worms, and their peculiar position, point to the possible belief that they may be signals, constituting a primitive means of communication; also of use to the animals in lighting their way, as we have seen in the case of the pyrophorus. The lights of fishes, whatever may have been the object of nature, serve several distinct purposes: to draw the attention of enemies, to attract prey, and to illumine the gloom about them. Any one who has fished at night by torchlight well knows the attraction that light has for fishes of all kinds, and when submarine electric lights have been watched, groups of fishes and squids have been observed

about them ; so it is evident that predatory fishes possessing
lights have in their lure a decided advantage.

Actual experiment has shown that the electric light can
be seen ninety-nine feet under water. The soft rays of
animal phosphorescence would not penetrate so far, but
would be powerful enough to illumine the water for some
distance about them.

The deep-sea fishes which are not remarkable for their phosphorescence, or do not possess it at all, have feelers in many
instances, and grope about like blind men ; while others have
eyes that not only see, but are possible emitters of light
themselves. In the case of the predatory shark captured by
Bennett, we may assume that the light was an effective lure :
but the same will not apply to the brilliant scopelus and
other delicate little creatures almost completely defenceless ;
so that it will be seen that it is as difficult to lay down fixed
rules for the use of the light as to explain the cause of its
production. The phosphorescence of corals and their allies,
— gorgonias, sea-anemones, etc., may serve to attract prey.
The minute crustaceans, so valuable to food fishes, are by
their unfortunate gift rendered visible to their enemies, and
the same may apply to many of the worms ; while in a certain
species of the genus *Polynœ*, we have seen that the phosphorescent scales which it throws off may be used to delude
its enemies, just as when certain lizards cast off their tails,
and dart away, leaving them wriggling and squirming, to
attract the attention of their pursuers. Certain crustaceans
have luminous bands or spots which undoubtedly serve as
lanterns, while many have eyes that are modified into light-
emitting organs. The light produced inadvertently by
schools of mackerel, in their movements through water teem-

PLATE XXV.

LUMINOUS WATERSPOUT.

ing with phosphorescent animals, redounds to the benefit of the fishermen. The pale phosphoric cloud, seen from the top masthead, resting upon the surface of the ocean, tells the secret of their exact situation; and, by surrounding it with the great net, large schools are often caught.

Among the insects we have definite experiments to show that the light they emit is a signal; in other words, the insects recognize the lights of their friends. A French naturalist one evening held from his window a living specimen of *Lampyris noctiluca* (Plate IX.) in the presence of several friends; and a few moments later a companion insect left the gleaming throng without, and alighted upon his hand, touching the captive, whose light was almost immediately extinguished.

M. Raphael Dubois, member of the Zoölogical Society of France, etc., has shown that the *Pyrophorus* (Plate XI.) uses its light as we would a lantern in the night. When he covered the light upon one side of the insect, it pursued a curved course; and, when both lights were extinguished, it was obviously at fault, and moved along with great care, and was evidently unfitted for nocturnal life.

We have seen how these insects were the means of saving the life of Jaeger, in lighting him out of the forests of the southern islands; how natives attach them to their feet, and employ them as lanterns; while others in South America form an article of trade, being utilized by the ladies as articles of personal adornment.

It must be evident to my young readers, that a practical application of the general features of phosphorescence would be extremely valuable, and in the previous chapter luminous paints and writing fluids have been referred to. An English chemist, named Balmain, has produced from Canton's phosphorus a paint which is luminous in the dark, and which has been applied to many purposes. Years ago the Chinese used a luminous paint made from powdered mussel-shells. The Emperor Tai Tsung, who reigned in the latter part of the tenth century, possessed a painting which, if examined by day, represented a cow browsing in an open pasture, but if this picture was taken into a darkened room, or looked at by night, the cow was seen to be lying down behind a fence, securely housed and protected. The secret was, that the fence and the cow in the night picture were painted in "South Sea pearl paste," as the Chinese called their phosphorescent paint, and were alone visible; while in the daylight the painting of "powdered reef-stone" only was seen, representing the animal in a standing position.

To Balmain, however, is due the credit of introducing luminous paint in this country and Europe, and it is applied to many objects. We have the faces of our clocks and watches luminous, so that the time can be told in the dark. Match-safes are rendered conspicuous by the same means, and various other articles.

Through the courtesy of Messrs. Devoe & Co., of New York, I was enabled to examine the application of this paint upon statuary and other objects. Upon entering a dark room, a statue was seen outlined in a wonderful bluish light of remarkable softness and beauty. An arm resting upon a table was vividly luminous, and presented a ghostly appearance. A large globe which hung from the ceiling gave out a soft radiance, quite sufficient to dispel the darkness, and the entire exhibition was suggestive of the varied uses to which the light could be put. Among these might be mentioned the painting of houses, so that they will render the streets luminous; buoys at sea; even the hulls of ships and their sails might be made conspicuous in this way. In London the harnesses of horses engaged in night work have been rendered luminous by this paint; and its availability in mines, and in large sewers like those of London, tunnels, and other subterranean works can hardly be estimated. Artificial fishes are painted, and used as luminous bait; and toys innumerable are placed upon the market, made interesting by application of this discovery.

It is obvious that luminous paint cannot be used in some cases, and to take its place Messrs. W. C. Home and E. Ormerod of London have recently invented a method of utilizing the luminous powder prepared mainly as a sulphide of calcium, for admixture with cements, plaster of Paris, and concrete, the object being to prepare the articles with a self-contained phosphorescent property instead of coating them with luminous paint. They take the proper proportion of any suitable cement, with the right amount of the luminous powder, mixing these with water, and moulding it to the required shape in the usual way, after which it is laid on the ceilings or walls with a trowel. The patentees attach importance to placing the moulded articles, as soon as dry, in a bath of paraffine wax and benzoline, or other waterproofing substance equally good.

In the case of using the luminous cement upon a wall or ceiling, they sponge or brush the surface over with a solution of paraffine wax and benzoline, or other suitable dampproofing solution. The uses of a luminous cement are manifold; e.g., for the garden, luminous concrete as edging to garden-paths and carriage-drives; for guides and beacons

at the entrance-gates of drives; insides of stables; the base
of balustrades, or the entirety of balustrades; for roads, as
luminous beacons of corners of dark country lanes, and at
the ends of bridges, ends of walls, and curbs of foot-paths;
for docks; for edging of piers and wharves; for water-
works; for the safety and despatch of night-work by the
erection of luminous guides and beacons; and for fire-plug
notices on walls; in short, for any place where the light of
day will sufficiently excite the phosphorescent property as to
render the cement or concrete work luminous by night.
The difficulty of sighting rifles in the dark has been ingen-
iously overcome by the use of luminous paint, and it is
thought that the armies of various nations will adopt phosphorescent sights for general use.

I have before me as I write, through the courtesy of
M. Raphael Dubois of Paris, a fine photograph of a bust
of Claude Bernard, taken by the light of numbers of phosphorescent insects (elaters), which shows the possibility of
work in this direction.

M. Ch. V. Zenger of Paris has made some interesting
experiments, and expressed the belief, some time in 1883,
that Mount Blanc could be photographed by phosphorescent
light emitted, and I understand this has been accomplished.
M. Zenger has photographed objects by the light of Balmain's
phosphoric plates. From a personal communication from this
scientist, I will quote some things which he has kindly sub-
mitted for the author's use in this volume, referring to this
work and the use of Balmain's liquid phosphorus. As a
light, he says, " No doubt there may exist better and more
perfect phosphorescent bodies of green, greenish blue, and
violet hue, than are at my disposal ; and to avoid the use of

sulphurets and sulphides, etc., and to obtain as long a phosphorescence as possible, is all I want to reduce stellar photography to the simplest and cheapest apparatus, and make it available to every one."

As we have seen, the light emitted by animals, plants, and minerals, of whatever cause, presents much that is mysterious; and the problem of animal phosphorescence would seem no nearer being solved to-day than it was fifty years ago. This is perhaps due to a lack of study and investigation. A glance at the appended bibliography shows that much has been written upon the subject; but it is only within the last decade that serious work in this direction has been done, typified in the superb work of Dubois, and the papers and monographs of the other scientists mentioned. The naturalists of the "Albatross," the government exploring-steamer, are to make investigations regarding the luminosity of the Pacific, during the forthcoming tour on the western coast. The French Academy of Sciences offers this year a prize of three thousand francs for the best paper upon animal phosphorescence. From this it would appear evident that the phenomenon is creating renewed or increasing interest, and in the following years will be the subject of much study and investigation; and we may expect in the near future to have not only its cause explained, but possibly to see a practical application of its possibilities to the wants of mankind.

APPENDIX.

1. PAGE 5. — *Noctiluca.* This interesting little creature belongs, in the natural arrangement as now recognized by science, to the first grand division of the animal kingdom. Simple as it is, it is not so completely without organs as some which form the first groups of this first division, as it has a whip-like organ, which gives name to its group, the *Flagellata,* or flagellate infusorians. These monads, as they are also called, are represented by a species of *Noctiluca* in our North-American waters off the coast of Maine. Huxley regards its luminous property as given out by the peripheral layer of protoplasm which lines the cuticle.

M. Giglioli of Bologna, Italy, in a letter to the author, says, "I have distinguished three modes of marine phosphorescence, very distinct, which present a great number of varieties. These are, —

" (a) Diffused homogeneous milky light.

" (b) Luminous points, sparkling and inconstant.

" (c) Luminous disks, with light generally fixed, and not sparkling.

" In one case the sea seemed on fire, and dolphins seemed to be fire.

" Again, the sea seemed to acquire an oily consistence, giving out soft homogeneous light, of a milky color, tinted with green or bluish. It is perhaps the least frequent, but most striking. It is due to the presence of noctiluca. It often resembled incandescent rain falling from the paddle-wheels of steamers."

M. Giglioli agrees with Huxley in stating that "the phenomenon of phosphorescence in these animals does not reside in the protoplasmic branches, which, as is known, are sometimes wanting; but in the cortical substance it is not uniform, but manifests itself in distinct and very minute luminous points, which sparkle, go out, and light up again."

169

2. PAGE 9. Kiel observed this phenomenon in *Peridinium.* The following species of luminous forms existing in the Baltic Sea have been described by Ehrenberg: *Prorocentrum micans, Peridinium michælis, Peridinium micans, Peridinum fuscus, Peridinium furca, Peridinium acuminatum, Lynchata baltica,* and a species of *Stentor.*

3. PAGE 9. Giglioli and his assistant, De Fillipi, observed luminosity in the gelatinous mass described by Hækel as *Citophora.*

The genera of those low forms most remarkable for luminosity are *Thalassicolla, Collozoum, Sphærozoum,* and *Collosphæra.* Giglioli states that the forms of this group which are found in the Indian Ocean and China seas are not luminous.

4. PAGE 11. *Dymophora fulgurans.*

5. PAGE 12. Other light-givers of this group are *Willsea prolifera, Bourganivillia,* and *Lizzia.*

6. PAGE 12. *Mueniopsis leidyii.*

7. PAGE 13. The *Lucernaria* is a very rare form of *medusa* on our northern shores, and particularly characteristic in color and form. It is more like a polyp in texture, and its rich beryl color distinguishes it from all other forms. It is related to the *Discophores,* animals belonging to one of the groups of jelly-fishes, or *medusæ.*

8. PAGE 14. Schafer has observed radiating fibres on the under side, but there is no evidence to show that the luminosity originates here. In fact, the outer surface, where the cells of the delicate epithelium, or skin, contain minute points of fatty material, is equally phosphorescent. The tentacles become luminous, and it is supposed that they contain no nerves except at the margin of the disk. In some instances the light seems well defined at the so-called eye-spots at the edge of the disk, but its sudden fluctuations render any attempt at locating a photogenic structure difficult.

While numerous theories are advanced, investigators are entirely at fault as regards any satisfactory explanation of the phenomenon. There are certain conditions which are not favorable to the emission of light; and observers have seen *medusæ* vividly luminous at one time, and not so at another.

It has been suggested that the light is subject to the so-called will of the creature. A better theory, perhaps, would attribute the luminosity to certain peculiar conditions, or to certain stages of existence.

9. PAGE 15. *Pleurobrachia rhodactyla,* Agassiz. This is one of the numerous free-swimming marine animals, belonging to the *Ctenophores.* A group of the sea-jellies which have the pretty rows of paddles adown their long diameter. They are usually about a pigeon's-egg in size, are oval, and in their element almost invisible, so colorless and transparent are they. A close inspection shows the paddles to be iridescent.

10. PAGE 15. *Idya roseola,* Agassiz. Another form found near the shores of Nahant.

11. PAGE 16. The *Physophoridæ* include the interesting forms, *Physalia* (Portuguese Man-of-War), *Porpita, Vellela,* etc. The first named indicates the character of the group, as its fleshy mass is surmounted by a beautiful bladder-like float, a mere bubble of membrane. These forms are not often seen out of tropical waters.

12. PAGE 16. The term zooids is applied to the mass of tentacles and other fleshy parts of the *Physophoræ.* The long, extensile feelers are for prehension ; others aid in locomotion, and some are reproductive; others are feeders for the entire colony. Thus it will be seen that these creatures are in a sense compound animals.

13. PAGE 24. Alcyonarian corals from an order in the class *Actinozoa.*

14. PAGE 24. Professor Moseley, of the "Challenger" expedition, was enabled to examine the light from these beautiful forms by the aid of the spectroscope, and found that it consisted of red, yellow, and green rays only.

15. PAGE 25. *Acanella normani Verril.* A pretty soft coral, which has been dredged off the New-England coast by the fish-commission. This is a revelation to science, as no one was ready to believe that such forms, so common to the tropical regions, would be found where they were. The Gulf Stream runs so close to the North-eastern States, it will not, on reflection, seem strange that some creatures common to the warmer waters may find a home there.

16. PAGE 25. *Primnoa resida.*

17. PAGE 25. *Paragorgia arborea.*

18. PAGE 26. *Pennatulidæ.* The name of a family of marine animals, which includes the *Umbellularias, Veretillum,* etc., — the last highly phosphorescent.

19. PAGE 26. While investigations so far have failed to explain the physiology of the light, it has been found that in a perfect animal it is emitted from eight opaque cords, each of which passes from a little swelling at the base of a tentacle down each polyp into the covering of the branch. The cords are canals in the sarcode of the branch, connecting the hollow of each tentacle with the tubular cavities of the branchlets and stem. The microscope shows that the contents of the canals are a fluid and cells ; the latter containing minute highly refracting globular particles of a fatty substance, which resists decomposition long after the death of the polyp itself. If these cords are ruptured, the luminosity of the entire mass is excited, and the fatty cell contents is luminous after its escape, and on foreign matter even after the death of the animal.

Regarding the light, Duncan says, referring to Panceri's experiments, " There is no sensible increase of temperature, and the tint of the monochromatic light is azure or greenish, but never red. In this beautiful instance of this remarkable vital luminousness there is evidently a photogenic structure and an elaborated organic material capable of producing light after removal from the animal. The sequence of illuminating the whole pen is slow, — far less than that of the movement of nerve-force. Yet the presence of the lowly organized nervous element indicates that the regulating of the light may relate to it as its function."

Perhaps the most magnificent of all the *Pennatulidæ* is the tall *Umbellularia graenlandica* (Plate XXI., Fig. 2), which consists of twelve huge polyps, each with eight fringed arms, terminating in a close cluster upon a stalk about four feet in height. This striking form was dredged by the " Challenger" expedition in water over two miles in depth, where the pressure is so great one can hardly realize it, and the temperature is just above freezing. Sir Wyville Thompson says, that, when this splendid animal was taken from the trawl, it emitted a light so brilliant that Capt. Maclear found it an easy matter to determine the character of the light by the spectroscope. It gave a very restrictedly continuous spectrum, sharply included between the lines *b* and *d*.

20. PAGE 27. *Pavonia quadrangularis.*
21. PAGE 27. *Asteronyx loveni.*
22. PAGE 27. *Ophiacantha.*
23. PAGE 28. *Renilla reniformis.*

24. PAGE 28. *Virgularia* is so named from its rod-like form; *vira*, a rod. *V. mirabilis* is found off the English coast.

25. PAGE 30. *Ophiura* and *Asterias.* These are genera of the sea-stars, or star-fishes long so called; the former so named on account of the resemblance to snakes in its arms.

26. PAGE 30. *Ophiothrix fragilis, Amphiura belli,* and *Ophiocantha spinulosa.*

27. PAGE 31. *Ophiocnida oliracca* and *Ophiocantha bidentata.*

28. PAGE 31. *Brisinga elegans.*

29. PAGE 32. *Astrophyton.* There are several species of this star-fish, but each found in deep water. They are curiously circumscribed in locality. In one place off Cape Cod they are dredged, but in no other place, excepting farther south. Their name, basket-fish, is from their numerous intwined arms, resembling basket-work.

30. PAGE 37. *Serpula.* A genus of the group *Annelida.*

31. PAGE 37. *Neiridæ* and *Eunicedæ.* Genera of the group *Annelida.*

32. PAGE 37. *Polynoidæ, Scyllidæ, Chætopteridæ,* and *Polycirus.*

33. PAGE 38. *Chætopterus norvegicus.*

34. PAGE 39. *Harmœthe imbracta* emits a bright greenish light when disturbed, the luminosity evidently proceeding from the point of attachment of each dorsal scale.

35. PAGE 40. *Pholas.* A clam-like mollusk. Several species are found on Nahant beaches. ' *P. dactylus* is a European form. The genus *Zirphœa* is found from New England to Great Britain. All are more or less borers. A small species bores in hard mud on the Nahant beaches. Others are known to bore into hard wood and into stone.

Pholas dactylus will be seen to have photogenic or light-emitting structures and substances almost concealed in the tissues of the animal. The light-emitting portions are, according to Panceri, "two parallel cords containing an opaque white matter extending down the anterior siphon, two very small spots at its entrance, and finally an arched cord corresponding to the superior edge of the mantle, reaching to the middle near the valves. The white color of the cords, which stand out in relief, distinguishes them; and, although they are only elevations of the sub-cuticular tissue, they contain special cells, or rather epithelium, which produces the phosphorescent matter. The whole surface of the Pholas is covered with ciliated epithelium, which dips down into all the parts of the animal; but the special epithelium differs from this. It is nucleated and crammed with granules, and the cells are very refractive. The cells are very fragile, and allow their contents — i.e., granular nuclei and refractive granules — to escape readily. These are soluble in ether and alcohol. Under ordinary circumstances this photogenic apparatus is hidden; but violence readily displaces the special cells, which burst, and their contents are carried all over the surface by the water, assisted by the general ciliation. The white substance, fat-like, retains its . luminosity, when spread out on paper, for hours; but the light does not appear to be accompanied by an evolution of heat. When it is placed in carbonic acid gas, the light pales and ceases. On the other hand, the photogenic substance, when barely luminous, is rendered so by physical contact. Agitation, and the addition of fresh or salt water, develop the light, and the same effect is produced by electricity and by heat. The light is monochromatic, and has a constant place in the spectrum as an azure band from E to F, that is to say, in the green."

36. PAGE 44. *Dendronotus arborescens.* A curiously decorated marine slug, found on the algæ of the waters around Massachusetts Bay. *Eolis* is another form nearly as interesting.

37. PAGE 55. The spectrum of the light of comparatively few of these beetles has been examined. That of *Photinus* was found by Professor C. A. Young, the astronomer, to be continuous without lines, and to extend from Fraunhofer's line C in the scarlet, to about F in the blue.

Mr. Meldola examined the spectrum of the light of the glowworm some years ago, and found that it was continuous, being rich in blue and green rays, and comparatively poor in red and yellow.

38. PAGE 56. Professor Carl Emery of the Entomological Society of Italy has kindly sent to us a detailed account of his experiments with the illuminating apparatus of a native luminous insect, the *Luciola italica*, etc. As these are the latest conclusions by the highest scientific authority, and therefore to be regarded as the most reliable, we here present a full account.

"The elytra of the insect *Luciola* were glued upon a holder of the microscope, and covered by a glass of tolerable thickness. On examining it, I got a favorable magnifying power, A of Zeiss. With stronger objective there is no good effect.

"The eye is at first dazzled by a strong, uniform yellowish light. But the intensity of this light is soon checked, the luminous field being interrupted by round spots. The light continues to diminish; the image becomes paler; and between the obscure round spots are seen to appear confused shadows, which detach themselves from the more brilliant rings. These rings are last to disappear when all the other portions have become dark. In the end they disappear entirely.

"The organ remains dark until the next flash; only here and there brilliant isolated points persist, which, as we shall see later, represent parenchymal cells which have retained their activity. If one places under the microscope the detached abdomen of a normal Luciola, and excites it by pressure of short duration by the cover glass, it is possible to obtain a flash which resembles the physiological flash."

M. Emery states that he found it unsatisfactory to examine the insect while alive, as the constant movements rendered it nearly impossible to observe correctly the phenomenon of luminosity. He proceeds: " I have found by poisoning the Luciola by vapors of osmic acid an excellent method in fixing the light, and studying exactly the microscopic aspect.

"When one examines in a dark chamber the abdomen detached from a Luciola which has been plunged in a solution of osmic acid, it is seen that a part of the segments occupied by the luminous organs shine with a feeble and variable light; whilst another part (ordinarily in the neighborhood of the median line) is obscure, or as it were veiled by a light phosphorescent cloud. When the preparation is placed under the microscope, the luminous parts exhibit towards the top the appearance which we have already noted in examining normal Luciolas; that is to say, the existence of obscure round spots surrounded by brilliant field. In observing more attentively, one perceives around the spots other little spots, less obscure, and sometimes hardly visible, disposed with a certain degree of regularity.

"Now, if we compare these images with those which are presented under the microscope by the luminous organs when hardened in alcohol, and cleared up by caustic potash, or else a tangetized section made of the organ of an animal killed by osmic acid, and colored by carmine, it becomes evident that the large, obscure round spots correspond to the central part of the digitiform lobes of *Targioni Tozzetti*; that is to say, to the cylinders constituted by the matrix of trachea (*Tracheenendzellen* of M. Schultze), whilst the luminous part is represented by the parenchymentous cells, and the little obscure spots are due to nuclei of these same cells. Still towards the limit of the brilliant and obscure regions of the luminous organ a very varied spectacle is observed. . . .

" From all the facts which we have just described, one may conclude with full certainty that the light of the Luciola has its seat in the parenchymentous cells of the luminous organ.

" It remains to be seen if the luminous combustion does not also take place, though, with luminosity in other parts. In my previous work I had it that the surface of the cylindrical lobes formed by the matrix of the tracheæ was the principal focus or seat of the combustion. The facts which result from later observation oblige me to abandon this opinion. . . .

" In the moments of mean luminous activity, one may say that the combustion is situated exclusively in the parenchymentous cells of the superficial layer of the luminous organ."

39. PAGE 73. *Gammarus caudisetus, Gammarus longicornis, Gammarus truncatus, Gammarus heteroclitus, Gammarus crassimanus. Cyclops exiliens* is also luminous.

40. PAGE 77. Another species in which this change had taken place is *Galathodes antonii*, an allied form which is shown in the central figure of the frontispiece. Many more, as *Willemœsea, Pentacheles, Polycheles*, and others, have organs of vision, which have undergone more or less change. It has been suggested that certain deep-sea crabs, as *Geryon tridens, Gonoplax, Donychus*, and *Munida*, have phosphorescent eyes.

In *Ptycogaster formosus* (Plate XIII., Fig. 1), we find an interesting form, living at a depth of twenty-eight hundred and fifty feet, or more than half a mile, from the surface, which is provided with well-developed eyes.

41. PAGE 81. The individual zooids, amounting to many hundreds, are grouped in whorls, their orifices so arranged that the inhalent are upon the outside of the cylinder, and the exhalent upon the interior. Each animal draws in a current from the outside, ejecting it into the interior ; the result of this volume of water rushing from the open end being that the entire colony is forced along, at the same time revolving upon its long axis.

42. PAGE 81. Panceri says, "Each zooid has two luminous spots, which are situated over the position of the ganglia of the nervous system; and there are loops like cords passing over the narrow end, connecting them."

43. PAGE 94. Dr. Gunther expressed the view that the organs are the producers, not the receivers, of light. He says, in brief, that the number of pairs of small globular bodies found along the abdominal profile is in direct relation to that of the vertebræ, the muscular system, etc. These are of two kinds. One class consists of the anterior, bi-convex, lens-like body, which is transparent during life; simple, or composed of rods, and coated with a dark membrane composed of hexagonal cells or rods arranged as in a retina. This structure characterizes the plates of *Stomias* (Plate XX.), *Astronechtes, Chauliodus* (Plate XXI., Fig. 4).

In the other set, as found in *Gonostoma, Myctoplum mausolicus,* and *Argyopelicus,* the organs have a simple, glandular structure. Branches of spinal nerves have been traced to each organ, and are distributed over the retina-like membrane of the glandular follicles.

The difference in structure of those organs naturally produces difference of opinion regarding their functions; but Gunther believes that all the organs in their functions have some relations to the conditions of light in which the fishes that possess them live. Three principal theories regarding them are given: first, they may all be accessory eyes; second, only the organs with the lenticular body are eyes, and those with glands are light-givers; third, all are producers of light. Many arguments have been advanced to support these different hypotheses; but it would seem that the second view is most tenable, from the fact that the organs with the retina-like membrane bear a great resemblance to a true eye, and finally the glandular organ in the little fish *Myctoplum* has been seen to gleam with a phosphorescent light. Dr. Gunther thinks it not improbable that the compound organ is an accessory eye, and a light-producer as well. The light, he says, may be produced at the bottom of the posterior chamber, and emitted through the lenticular body in particular directions, with the same effect as when light is sent through the convex glass of a bull's-eye.

44. PAGE 104. *Orthogoriscus mola.*

45. PAGE 128. Diatoms: *Pyrocystis pseudo-noctiluca* and *P. fusci-formis.*

46. PAGE 138. The common mushroom (*Agaricus campestris*), the meadow mushroom (*Agaricus arvensis*), the French champignon (*Marasmius oreades*), and in Austria *Agaricus mellius*, are eaten largely. Truffles (*Tuber æstivum*), Morels (*Morchella esculenta*), and Puff-ball (*Lycoperdon giganteum*) are also favorites in Europe.

47. PAGE 138. Jew's ear: *Himcola auricula Judæ.*

48. PAGE 138. *Myletta austratis.*

49. PAGE 149. See the description of ooze on page 3. This ooze is formed of the cast-off shells of the *Diatoms*, the minute vegetable forms of low organization.

BIBLIOGRAPHY

OF

WORKS ON PHOSPHORESCENCE.

A.D.

1526. Oviedo y Valdes Gonzalo Fernandez. De Luminario de la Natural y General Istoria de las Indias. Toledo, 1526.

1536. Anglicera (Pietro Martine d'). Decades of the New World (De rebus Oceanio et Orbe novo Decades). Paris, 1536.

1634. Moufet. Insectorum sive Minimorum Animalium Theatrum.

1635. Joannes Eusebius Nierembergius. Hist. nat., lib. xiii. c. iii.

1647. Bartolin, Thomas. De Luce Animalium, lib. iii. Lugd., Batav., ex officina Francisi Hackii, p. 205.

1667. Dutertre. Histoire générale des Antiles françaises, p. 280. Paris.

1667. Stubbes. A Continuation of the Voyage to Jamaica. Philosoph. Trans. No. 36 (Mem. of Royal Society, 2d ed. 1745). Id., Jour. d. Scav., 1667.

1668. Norwood. Observations in Jamaica. Philosoph. Trans. No. 41 (Mem. of the Royal Society, i. London).

1725. Sloane. A Voyage to the Islands Madeira, Barbados, Nieves, St. Christophers, and Jamaica, etc., ii. p. 206. London.

1742. Melchior, Johan Alb. De Noctilucis (Lampyris, Elater). Franequeræ. Dissert. philosoph. (Bibl. de Lacordaire).

1756. Brown. The Civil and Natural History of Jamaica. London.

1763. Gronov. Zoöphylacium Gronovianum. Lug., Batav., p. 152, No. 474.

1764. Linné. Museum S. R. M. Lud, Ulr. Reg. Holmiæ, 1764, p. 83, et Syst. Nat. ed. 12, tome i. part ii. Holmiæ, 1767, p. 651.

1766. Fougeroux de Boudaroy. Mémoire sur un Insecte de Cayenne appelé Maréchal et sur la Lumière qu'il donne. Mém. Acad. d. Sc., p. 339.

1774. De Geer. Mémoires pour servir à l'Histoire des Insectes. Stockholm. iv. 1774, pp. 160, 161.

1790. Olivier. Entom. ii., p. 15.

1805. Palisot de Beauvois. Insectes d'Afrique, et d'Amérique. Paris.
1807. Illiger. Monographie der Elateren (Elateren mit leuchtenden Flecken
      auf dem Halsschilde).
1807. Viviani. Phosphorescentia Maris, quatuordecem novis speciebus illus-
      trata. Genua, 1807.
1809. Azara. Voyage dans l'Amérique méridionale, 1781-1801, i. p. 114.
      Paris, 1809. 4 vol. in 8° et Atlas.
1810. Macartney.   Observations upon Luminous Animals.   Phil. Trans.,
      pp. 277, 279, v. 100.
1814. Humboldt et Bonpland.   Voyage au Nouveau Continent, iii., p. 482.
      Humboldt, Relation historique, i. p. 79 et 533.   Humboldt, Ta-
      bleaux de la Nature, ii. p. 69. Paris, 1851.
1817. Gilbert, M.   Annales Maritimes.   1817.
1817. Kirby and Spence.   Introduction to Entomology, p. 513, et Note,
      abrigée de Morren, 7e éd., 1856, p. 503 et suiv.
1823. Spix et Martius. Travels in Brazil, 1817-1820.
1827. Curtis.   Account of Elater noctilucus of the West Indies.   Zoöl.
      Journ. iii. p. 379.
1830. Lacordaire.   Mémoire sur les Habitudes des Insectes Coleoptères
      de l'Amérique méridionale.   Ann. d. Sci. Nat. xx. p. 241, et Intro-
      duction à l'Entomologie.
1832. Burmeister.   Handbuch der Entomologie, i. p. 535.
1832. Latreille. Voyage de Humboldt, Recueil d'Observations de Tropiques
      dans les Années 1799-1804, v. Paris, 1811-1832. Insectes, i. pp.
      127-304, pl. xv. à xxv., ii. p. 9 à 138.
1838. Dugés.   Traité de Physiologie comparée, tome iii. Montpellier, 1838.
1840. Tessan, M. de.   Comptes Rendus de l'Académie des Sciences. 1840.
      Rapport fait par M. Arago.
1840. Hooker's Journal of Botany.   Light of Agarics.   1840. p. 426.
1841. Coldstream.   Todd's Cyclopædia of Anatomy and Physiology, part
      xxii. 1841. Article on "Animal Luminousness."
1841. Heward, Robert.   Memoir of the Fireflies of Jamaica.   (Verschie-
      denheit in Emission des Lichtes bei Elater und Lampyris.)   The
      Entomologist, pp. 42-43.
1841. Morren.   Sulla Fosforescenza delle Lampiridi noctiluca e splendidula;
      Atti terza Riunione Scienz. Ital., Firenze, 1841, p. 366. Isis, vii.
      p. 412, 1843.
1841. Germar.   Beiträge zu einer Monographie der Gattung Pyrophorus.
      Germar's Zeits. d. Entom. iii. p. 1, 1841. Bemerkungen über
      Elateriden. Germar's Zeits. Entom. iv. p. 43, 1843; v. p. 133,
      1844.
1841. Erichson.   Of Pyrophorus from Cuba (sic).   Wilgmann's Archiv, i.
      p. 87.

1843. Dowden. Proceedings of the British Association. Light of the Marigold.

1843. Lankester, Dr. E. Luminosity of the Marigold. Gardener's Chronicle, 1843, p. 691.

1844. Reiche. Note sur les Propriétés lumineuses du Pyrophorus nyctophanes. Ann. de la Soc. ent. franç. (2), ii., Bull., pp. 63–67.

1845. Luminous Vegetables. Proc. Royal Asiatic Society. April, 1845.

1848. Tulasue. Light of Agarics. Annales des Sci. Nat. Paris, 1848. ix. p. 340.

1848. Tulasue. On the Phosphorescence of Vegetables. Ann. des. Sci. Nat., 1848, vol. ix. 340.

1850. Burnett. On the Luminous Spots of the Great Firefly of Cuba (Pyrophorus phosphorus). Proceed. Boston Soc. Nat. Hist. iii. pp. 290–291.

1853. De Lacaze-Duthiers. Recherches sur l'Armure génitale femelle des Insectes. Ann. d. Sc. Nat. (3), xix. pl. 3, figs. 6, 7, 8, 9, 1853.

1853. Reinhardt, J. F. Twende Jagttagelser af phosphorish Lysning hos en firk og ep Insectlarve. Vidensk Meddel. fra. d. Naturalist. Foren Kjœbenh. for 1854, pp. 60–65. Transact. Entom. Soc., London (2), iii. 1854 (Proceed., pp. 5–8). Zeitschrift f. d. gesammt., Naturw, v. 1855, pp. 208–212.

1855. Van der Hoeven. Einige woorden over het Lichten van den Zuid-Amerika anschen Springkever. Album der Natur, 1855; aflev., 7, pp. 205–212.

1859. Light of Scarlet Verbena. Gardener's Chronicle, July 6, 1859, p. 604.

1860. Montronzier. Essai sur la Faune entomologique de la Nouvelle Calédonie. Ann. Soc. Entom. Fr., p. 258.

1863. Dr. Hahn. Luminosity of Dictamnus Albus. Journal of Botany, 1863.

1863-65. Caudèze. Monographie des Élatérides; Mém. de la Soc. Roy. d. Sci. de Liège, xvii. pp. 1–70. 1863. Sep. iv. Élatérides vrais, p. 70, pl. i. fig. 3 et fig. 5 à 23.

   Id., Élatérides nouveaux. Mém. couron. de l'Acad. Roy. de Belgique, xvii. p. 51. 1865.

1864. Pasteur. Sur la Lumière émise par les Cucujos. Compt. rend., Acad. d. Sc. (2), lix. No. 12, p. 509.

1864. Blanchard (E.). Compt. rend. lix. p. 510, 1864. Métamorphoses, Mœurs et Instincts des Insectes, p. 537. Paris, 1868. Compt. rend. lxxvii., p. 333, 1873.

1865. Milne-Edwards. Leçons d'Anatomie et de Physiologie, viii. leç. 68ᵉ. Paris.

1867. Perkins, G. A. Bericht den "Cucujo" oder Westindischen Leuchtkäfer (Elater noctilucus). Am. Natur., viii. pp. 442, 443.

1868. Murray, Andrew.   On an Undescribed Light-giving Coleopter-larva. Soc. Linn., p. 74, pl. i.

1868-74. Thompson, C. Wyville.   The Depths of the Sea.   Account of the general results of the dredging-cruises of H.M.SS. "Porcupine" and "Lightning." 1868, 1870, 1874. London.

1869. Smith.   Larve von Pyrophorus Uruguay. Proc. Ent., London, p. xv. 1869.

1870. Phipson.   Phosphorescence; or, The Emission of Light by Minerals, Plants, and Animals. London, p. 146 et suiv.

1871. Burmeister.   Käfer-larve von Parana.   Proc. Linn. Soc., xi. p. 416*ff*.

1871. Leprieur.   Soc. entom. de France, Bull., p. 68, à propos d'un passage du voyage à la Nouvelle-Grenade du Dr. Saffray, Tour du Monde, 605ᵉ liv.

1871. Light of Tuberose.   Science Gossip, 1871, p. 122.

1872. Fox-fire.   Gardener's Chronicle, 1872, p. 1258.

1872. Heinmann, Carl.   Untersuchungen über die Leuchtorgane der bei Vera Cruz vorkommenden Leuchtkäffer. · Erste Mittheilung Arch. f. mikrosk. Anat., viii. p. 461.

1873. De Dos Hermanas.   Sur las Cucujos de Cuba.   Compt. rend., tome lxxvii. p. 333.

1873. Gerard.   Les Taupes Lumineux.   La Nature, 1ʳᵉ année, p. 337.

1873. Robin et Laboulbéne.   Appariel lumineux des Cucujos.   Compt. rend. Acad. d. sci. lxxii. p. 511.

1874. Torrend Glover.   Report of Entomologist, and Curator of Museum, pp. 152-169, fig. 1 à 10. Habits and Luminosity of Pyrophorus physoderus (fig. 3), compared with P. nocticulus (fig. 4), and Photinus pyralis. K: A luminous Elaterid? larva.   Psyche, v. 1.

1874. Beach, A.   The Science Record for 1874.   A Compendium of Scientific Progress and Discovery during the past year, with illustrations, viii. p. 598, avec fig. New York.

1875. Darwin.   Voyage d'un Naturaliste autour du Monde.   Paris.

1875. Pickman Mann's Note on the Luminous Larvæ of Elateridæ (Asaphes memnonius). Psyche, i. p. 89.

1876. Richard Napp.   Cours sur les Arthropodes de la Faune de la République Argentine (Lampyrides et Élatérides, p. Weyenbergh). Die Argent. Republik.

1876-77. Wyenbergh, II.   Eine leuchtende Käfer-larve.   Horæ Soc. Rossicæ, xii. p. 177. xii, p. 177. fig., tab. iv. B.

1881. Gadeau de Kerville, Henri.   Les Insectes phosphorescents.   Rouen.

1881. Caudeze.   Élatérides Nouveaux.   Mém. Soc. d. Sc. de Liège, ix. (2).

1882. Bowles.   On Luminous Insects.   Rep. Entom. Soc., Ontario, pp. 34-37, fig. 16 (figure rep. a Pyrophore).

1883. Macgowan, Dr. Antiquity of Luminous Paint, and Chinese Method of making it. Science, 1883, ii. p. 698.

1884. Becquerel, M. Traité de Physique comparée dans ses Rapports avec la Chimie et les Sciences Naturelles, ii. 1884.

1884. Dubois, Raphaël. Propriété physique de la Lumière des Pyrophores (en commun avec M. Aubert). Acad. des Sci. 1884.

1884. Dubois, Raphaël. Sur la Lumière des Pyrophores. Soc. de Biol. 1884.

1884. Dubois, Raphaël. Note sur la Physiologie des Pyrophores. Soc. de Biol. 1884.

1884. Dubois, Raphaël. Note sur la Phosphorescence des Poissons. Soc. de Biol. 1884.

1884. Dubois (R.) et Aubert. Sur la Lumière des Pyrophores. Compt. rend. Acad. d. sc. Paris, 1884.

1884. Dubois, R. Note sur la Physiologie des Pyrophores. Soc. de Biol., (8), i. No. 40. Paris.

1885. Dubois, R. Fonction photogénique des Pyrophores. Soc. de Biol., (8), ii. No. 30, p. 559. Paris.

1885. Dubois, Raphaël. Note sur l'Action des hautes Pressions sur la Fonction photogénique du Lampyre (en commun avec M. Regnard). Soc. de Biol. 1885.

1886. Dubois, Raphaël. De l'Action de la Lumière émise par les Êtres vivants sur la Rétine et sur les Plaques du gelatine bromure. Soc. de Biol. 1886.

1886. Dubois, Raphaël. Les Élatérides lumineux. Meulan, 1886.

1886. Filhol, M. H. La Vie au Fond des Mers. 96 figures dans le texte et 8 planches hors texte, dont 4 en couleurs.

1886. Zengler, Ch. V. La Phosphora-graphie appliquée à la Photographie de l'Invisible. Paris. Acad. Sci., August, 1886.

1886. Zengler, Ch. V. Études photographiques pour la Reproduction photographique du Ciel. Comptes Rendus, No. 8, Feb. 22, 1886.

1887. Heineman, Ralph. Pyrophores. Vera Cruz, 1887.

Birard, M. Cited by Dugés, Traité de Physiologie, tome ii.

Carus. Traité élémentaire d'Anatomie comparée, traduit par Jourdan, tome i. ·

Cooke, M. C. Fungi, their Nature and Uses, p. 105.

Ehrenberg. Das Leuchten des Meeres. Abhandlung.

Gosse. Insects of Jamaica. Ueber das Leuchten von Pyrophorus noctilucus. Ann. and Magazine Nat. Hist. (2), i. p. 200.

Journal of the Linnæan Society, vol. x. p. 469.

Josephus. Luminous Roots. Wars of the Jews, book vii. chap. vi.

Leydig, Professor. Bonn, Germany. Phosphorescence of Fishes. 10 plates.

Lesson. Dict. des Sc. Nat. Article "Phosphorescence."

Macaire. Journal de Physique, tome xciii.

Matteucci. Leçons sur les Phénomènes physiques des Corps vivants. S⁰
    leçon.

Quatrefages, M. A. de. Silliman's Am. Journal of Science.

Tingry. De la Phos. des Corps, et particulièrement de celle des Caux de la
    Mer. Journal de Physique, tome xlvii.

Ussow, Dr. M. University of St. Petersburg, Russia. The Eye-like Organs
    of Fishes.

Van Benedin. Recherches sur la Cause de la Phosphorescence de la Mer
    dans les Parages d'Ostende. Bulletin de l'Académie Royal de
    Belgique, tome xiii., par. 2, p. 3.

INDEX.

www.ingramcontent.com/pod-product-compliance
Lightning Source LLC
Chambersburg PA
CBHW020847270326
41928CB00006B/584